An Introduction
to Models in Geography

An Introduction
to Models in Geography

Roger Minshull

LONGMAN
LONDON and NEW YORK

LONGMAN GROUP LIMITED
Burnt Mill,
Harlow,
Essex CM20 2JE

Distributed in the United States of America by Longman Inc., New York

Associated companies, branches and representatives throughout the world

© Longman Group Limited, 1975

First published, 1975

ISBN 0 582 48109 0 *cased*
ISBN 0 582 48108 2 *paper*
Library of Congress Catalog Card Number: 74—84344

Set in IBM Century, 10 on 11pt

Printed in Great Britain by
Whitstable Litho Ltd

Contents

Figures

1
The need for models

In the middle of the twentieth century there was a geography teacher who, because he was young, inexperienced and under-paid, was naturally expected to teach all branches of geography with equal knowledge and skill. His lectures on landforms, cyclones and soils were magnificent. He would enter the class-room looking forward to explaining a cycle of erosion or the process of leaching. His logical exposition was a delight to follow, and his multicoloured diagrams were a joy to behold. Many of his pupils, the girls included, were inspired to become geologists, meteorologists and even soil scientists.

His lectures on the farming, manufacturing and the distribu-tion of population in different parts of the world were in-describably dull. Although he believed this part of geography to be the more interesting and important, his pupils endured the boredom, and went away convinced that human geography was just a matter of learning one damned fact after another. There was no logical exposition, there were no systems to illustrate by diagram, little to give any kind of intellectual stimulation and satisfaction.

Imagine, in another time and another place, this fictional geographer, perhaps older and shrewder, has specialised in the branch of human geography which preoccupies him most. For many years he has been dictating the facts of urban growth to students who couldn't care less: between 1951 and 1961 Manchester grew by so many thousand, Bradford by so many, and so on. By a mistake on the part of his employers he gets some time off to think, and he begins to wonder why these towns are growing, at exactly what rate they are growing, and whether all towns are growing at the same rate.

So the research geographer has a problem or question about the distribution and movement of people over a particular part of the earth's surface. His problem contains certain elements which will be essential to the discussion of models to be pursued in this book. Namely, he has certain phenomena, in this case the individuals of the population and the cities of different sizes; secondly there is some movement or interaction between the phenomena, in this case migration of people to cities. In fact he has tentatively identified what he believes to be a system on the earth's surface, and his aim is to describe and explain the functioning of that system as accurately as he can.

This geographer now follows the ideal course of action. Having observed some of the facts of a very complicated system, which is made infinitely more complicated by being inextricably interconnected with countless other systems, he proceeds to set up an hypothesis about the structure of the system and the way it works. Together with a *verbal* description and explanation, this hypothesis may well take the form of a diagram to illustrate the structure, and a mathematical equation to represent the functioning.

For the moment, being interested only in the growth of towns, the elements of the system are defined as the native populations of the cities and the non-city population of the rest of the country. The growth of any city's population is then hypothesised as a function of both the size of the native population and the strength of the attraction of that city for the non-city population surrounding it. The geographer may decide to simplify and isolate further, in order to describe and understand one particular aspect of this. For argument's sake, he puts forward the hypothesis that the size of the migration to cities depends directly on the sizes of the cities; in other words, that a city of 1 million inhabitants attracts twice as many immigrants as a city of 500 000 inhabitants over the same period of time.

A diagram may be redundant in this example, but two things are essential: first, a mathematical equation to express the magnitude of migration to towns of different sizes, and, secondly, a suggested or hypothetical explanation of *why* people behave in this manner. Such an explanation might run along the lines that many more people are attracted to the city of 1 million rather than the city of 500 000 because the former offers a greater variety of employment, a better chance of housing, a greater variety of shops and services, possesses colleges, theatres and other amenities which the other does not possess.

With his hypothetical diagram, equation and explanation, our geographer has created a model of the real-life system he is studying. The model is a simplified hypothetical description and explanation of the interaction of phenomena on the earth's surface, and because the geographer is following the ideal course of action, his one aim in life is now to test his model, to test his hypothesis, until he proves it wrong. If this story were true, of course, he would simply select evidence to show that his hypothesis was a true description and explanation of what actually happens in real life!

However, matters have been made worse for him by the fact that another geographer, in another university, has observed exactly the same phenomena but has constructed a different model. This rival puts forward the hypotheses that the main movement is from one city to another, and that the medium-sized cities are the ones which are growing fastest. This second worker in the field discounts the importance of the movement from non-city to city, and expects to find that cities of 750 000 are the main attraction, because those of 500 000 offer too few amenities, while those of a million and more are overcrowded and congested. His diagram of direction of flow, his equation, and his suggested verbal explanation, are, of course, different from the first man's. In short, each man has set up a different model as a tentative, *simplified* description and explanation of the same, partially investigated, complex phenomenon.

The rival models must both be tested rigorously. It is not that one will be right and one wrong: one of the models may, certainly, prove to be on the right lines, but both could be wrong, each could contain part of the truth, and at best, if one seems to be right, it will be only an oversimplified start at the truth. It turns out that the rival geographer is as meticulous as the first, and both proceed to test their models rigorously and correctly. By now they are both experienced men, and give only a tight little smile when their colleagues babble on about the wealth of quantitative data just lying about waiting to be analysed. They know that there is perhaps one chance in a thousand that the data available on migration between cities, and to cities, will be complete, up-to-date, and in exactly the form in which they will need it.

So they, and perhaps their research assistants, start to collect the data they need, and to devise tests which will give valid answers. So we can imagine that our geographer tests his model ruthlessly and has to modify it in the light of new information, but that basically the model survives. Having dealt with the factual descriptive data, he then spends twice as long devising, getting completed, and analysing questionnaires to test whether his suggested explanations are also anywhere near the truth. They are, and eventually his model is so well tested and proved that he can publish it as one tiny part of a future theory of urban geography.

This attempt to outline what is meant by a model before going into more precise detail may exasperate those who are familiar with models. Much has been left out in this

I The elements and the structure

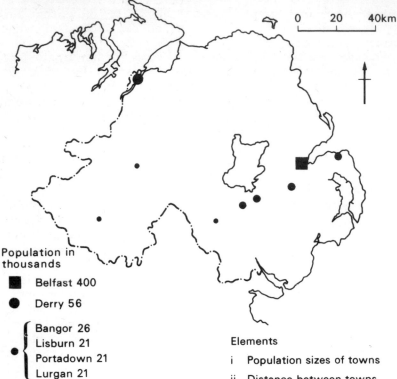

0   20   40km

Population in thousands

■ Belfast 400

● Derry 56

● {
Bangor 26
Lisburn 21
Portadown 21
Lurgan 21
}

● {
Armagh 10
Omagh 8
Enniskillen 8
}

Elements

i   Population sizes of towns

ii   Distance between towns

iii   Goods to be transported

II The functioning of the system

$$T i j = \frac{P i \, P j}{d i j^2}$$

Where

T = The amount of freight moved between any two of the towns designated i and j for each calculation

Pi = Population size of town i Pj = Pop size of town j

dij^2 = Distance between i and j squared

The *movement* of goods. people, information etc. is represented by the gravity formula

III The hypothetical explanation

That larger towns produce and attract more freight than small towns
That more freight moves between large towns and between towns close together, than between small towns and towns far apart.
That the volume of transactions between towns depend on their sizes and distances apart.

Fig. 1.1   The three parts of a complete model

simplification; only one type of model has been illustrated, and the many implications have not been followed up. As most of the chapters which follow deal with rather fine distinctions of the types, uses and misuses of models and closely related techniques, it was considered essential to give a working idea of the concept of the model at the beginning, so that this rough and ready idea can be improved and perfected in succeeding pages. In addition to this initial statement of the nature of a model, it is hoped that the narrative introduction has suggested certain assumptions which I have made, and from which my statements about models necessarily follow.

These assumptions are five in number. First, that models are already familiar tools in physical geography, even at school level, though as yet not necessarily under the label of model. W. M. Davis's cycle of erosion, and concept of youth, maturity and old age are models. That some geographers jeer at the cycle concept now, and others would call the concept of youth, maturity and old age an analogy rather than a model, only strengthens points to be made in this book. Models accepted as valid in one decade have to be either modified or replaced completely as new evidence is found, and as better hypotheses are put forward. The fact remains that much model-based teaching in physical geography has been going on for a long time.

Second, that models originate essentially as a tool for research. This important point tends to be lost sight of simply because so many tested and accepted models *do* end up as superb media for teaching. Third, and very closely connected with this origin in research, is the point that it is a geographer's attempt to describe and explain something which concerns him that creates the need for the model. The model is set up as one of the many possible descriptions and explanations of part of reality. Only exhaustive testing will help decide whether it is anywhere near the truth. The author rejects out of hand the suggestion that models have resulted from the need to deal with the mass of quantitative data which some geographers believe is now available. He has yet to see his data piling up round his shoulders, particularly the kind of data he needs, and he will not subscribe to the implication that the geographer is just an automaton which reacts to a stimulus. If geographers were standing around with their minds in neutral until the data came along, then he would rather be a bus driver. A research question sometimes raises the need for a model as the appropriate tool, and this in turn sometimes calls for specific numerical data to test it; then, and only then, are certain quantitative techniques necessary and relevant to geography.

Fourth, while there will be no attempt in this book to urge

research workers, lecturers or teachers in human geography to adopt a model-based approach, and certainly not to reject everything they have ever used or known in favour of fashionable models, it will be contended that models can be as useful in human geography as in physical, and that their use where relevant, and with restraint, can help to clarify the aims of research and can provide absorbing concepts in teaching. At all levels there is a need for logical explanation of matters of vital concern in human geography. Models are not the sudden, final solution this way, but their increasing use can help to give a formal, logical framework for more accurate explanation of those things we consider to be important.

The fifth assumption must be about the nature of geography. I have argued elsewhere that the distinguishing characteristic of geography is its study of the spatial arrangement of certain phenomena on the earth's surface (Minshull, 1970).* However, being well aware that there are many other views of the nature of geography, it would be pedantic and misleading to examine models only as they relate to one concept of geography. While the intention at one stage was to confine this study only to overtly spatial models, this has been rejected. This may seem to be a wide, unstructured view of the nature of geography, but it results from an attempt to deal objectively with all the different statements about models in geography which have come to my attention. In a later chapter the relevance of models to five distinct concepts of the nature of geography will be examined, but for the moment, no one view of the nature of geography will be preferred. For example, I do not accept that geomorphology is geography, but so many people accept it as part of physical geography, that this will be taken for granted in what follows.

The most comprehensive volume on the subject as yet available is *Models in Geography* (Madingley Lectures), edited by Chorley and Haggett (1967), in which each contribution deals with a specific aspect (see Bibliography, p. 155). Brief consideration of these essays may help to illustrate what is meant by models, show how they help to clarify the aim of each branch, and point up the differences in the position of models in physical and human branches of geography. What follows is necessarily very condensed, with many omissions, and those concerned with models are urged to study *Models in Geography* (hereafter cited as *MG*) very carefully.

Writing of hydrological studies, More (*MG*, ch. 5) states that the basic system which needs to be modelled is the hydrological

---

* Details of references are given in the Bibliography, p. 155.

6

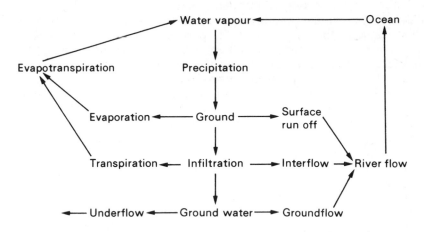

Fig. 1.2   A model of part of the hydrological cycle

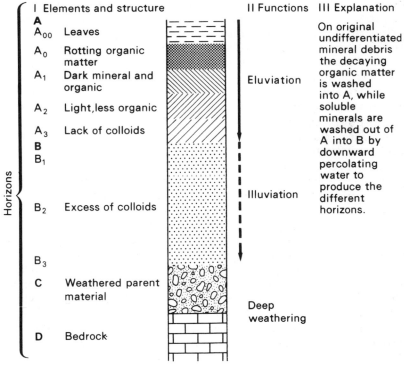

Fig. 1.3   A model of a soil profile

cycle. The reader may already be familiar with an early model of this system: water evaporating from the sea, being blown inland by the winds, raining down on to the land and returning to the sea and the air by runoff, percolation and evaporation. Later models, of course, have been constructed in an attempt to explain more of the details, say the events and processes in one river basin; the diagrams are more elaborate, the explanation is more detailed and refined, and, above all, there are now the mathematical elements of the model to make precise statements about such things as the rate of evaporation, the amount of water received by a given area of land, the proportions of runoff, evaporation and percolation, and so on. In short, qualitative statements are being replaced by quantitative measurements.

But the aim remains the same. There may be more models, in more detail, revealing a much more complex situation than at first expected, but the hydrologist still knows exactly what system he is trying to model, describe and explain. He aims first of all to understand the system, then to predict how it will operate in the future, say, to predict the rate of runoff or flood for a given rainfall, and thirdly, on the basis of this knowledge, eventually to plan the best uses of water resources.

Barry (ch. 4, on models in climatology and meteorology) writes of three types of models; those of the general circulation of the atmosphere, those synoptic models of such things as depressions and the monsoon system, and 'models characterising various aspects of the complex processes in the atmosphere'. Again we are familiar with some of the simpler models; but in this case we can not say earlier models. While the earliest model of a hydrological cycle mentioned by More is Horton's of 1933, the earliest meteorological models date from those of Halley in 1686 and Hadley in 1735. Similarly, Chorley (ch. 3) applies the word model to work as early as 1918.

However, the comparatively recent models of the circulation of the atmosphere and the operation of the monsoon which we learned at school are now hopelessly out of date. Bluntly, the three-cell model of the trade winds, westerlies and polar winds is too simple a concept of the real circulation, while the currently accepted model of the monsoon system contains the influence of the global wind system as a factor just as important as the annual high and low pressure systems over Asia triggering off the monsoon winds. The current model of any wind system would be unfamiliar to those of us who have not bothered with such things since out undergraduate days in that the diagram would be different, there would be many equations to quantify the movement of air, and the verbal explanation, while uncannily similar in some respects, would make very nice distinctions about what forces caused what movements of air. The references to conservation of energy, thermodynamics and hydrodynamics might frighten us back to rural settlement straight away.

Yet the climatologist, with these new models, these new versions of a very old tool, is still trying to do what he always tried to do, to understand the circulation of the atmosphere and the resultant climates by means of making a simple model which is both within his powers to comprehend and to test objectively. The circulation of the atmosphere needs to be modelled if only because it is invisible, but the model contains more than this. It contains a description in words, maps, diagrams and mathematics of the state of the atmosphere at one point in time; these are the elements or components of the system. It contains a description, in similar verbal, graphic and mathematical terms of the movements of air; this is the functioning of the system. It ought to contain, moreover, at least a tentative explanation of why the system functions or operates in this way. However much the contemporary climatologist laughs at the earlier models, and however seriously he takes the current one, the latter is only one more hypothesis about the nature of the atmosphere, and will be acceptable only until it is disproved and a new, more refined model has to be built. Only the quest remains constant, the desire to understand the atmosphere, and to be able to forecast events, the weather, with greater and greater accuracy.

According to Chorley (*MG*, pp. 59—61) geomorphology is precisely at the stage where one model has been found to be inadequate and a satisfactory new one has yet to be built. So much has been 'explained' by the models suggested by Davis that, now he seems to have been discredited in certain respects, everything Davisian is avoided, the good thrown out with the bad, and a new hypothesis is needed. Chorley suggests that in this search for a new hypothesis, each country is now developing different methods and aims and he himself suggests four different problems which geomorphologists might investigate:

1. The cause and effect in time of the processes which produce given landforms;
2. The spatial relationships, e.g., of structural provinces or assemblages or types of landforms;
3. The functioning of a particular system, e.g., the drainage basin, or desert erosion, transport and deposition;
4. The development of individual landforms.

The most striking point to emerge from the fact that geomorphology is without a satisfactory model at the moment, and

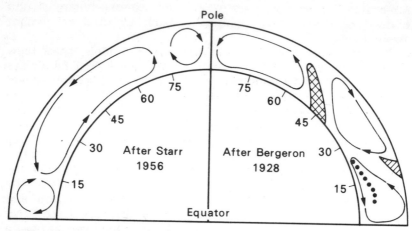

(a) **New** and old models of the general meridional circulation

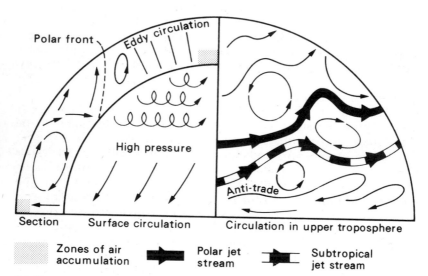

(b) **Model of tropospheric circulation in N. hemisphere, by Birot after Palmén, giving just a hint of the increasing complexity of such models as they get nearer the truth**

Fig. 1.4   Models of atmospheric circulation

from a comparison with hydrology and climatology, is that geomorphologists (now that they have disowned their leader Davis) appear not to know exactly what they are trying to explain. No amount of borrowing of ready-made models from other disciplines is going to be any use in the long run if, as Chorley suggests, this is the case. The procedure must be (*a*) this is what we want to understand, (*b*) here is a model put forward as a suggested explanation of it, and *not*, as is sometimes advocated (*a*) here is a nice model that meteorologists have a lot of fun with, (*b*) now how shall we use it? what shall we explain? at what tangent shall we fly off today?

Part of the reason for the geomorphologist's present confusion helps to throw more light on the nature of models. Both the hydrologist and the meteorologist are basically interested in systems; the system of water movement and the system of air movement. In contrast, it seems to me that most geomorphologists up to now have been interested in objects, in landforms as they are now, and in systems of erosion, transport and deposition only in so far as they help to explain the present landforms. So the former two disciplines have the emphasis on flow, process, function, with few concrete elements or phenomena, while the latter discipline has the emphasis too strongly on elements and not enough on the processes and interrelationships between them. This is not a criticism of geomorphology, simply a statement that it is easier to construct a model of a system when both the elements and the functions of the system can at least be identified.

If we put the emphasis on river, ice, wind and sea action in geomorphology, then this discipline, as well as climatology, pedology and biogeography all appear to study dynamic systems. By study of dynamic systems is meant here study of the interaction of various phenomena on the earth's surface to produce periodic or progressive change in the spatial distribution of those phenomena. The students concerned are not simply aiming to describe or catalogue, nor to explain how a certain static state of affairs has come about; they are aiming to describe and explain a continuing interaction of forces and objects. One can, therefore, discern a basic similarity in the models they use, and in all the physical disciplines there is the hope and belief in the possibility of prediction of events and control of the forces in question.

Now just as one would not want geomorphologists to adopt a method used by a climatologist merely because it had been successful in climatology, one must be wary of advocating that human geographers should follow blindly the techniques of physical geographers. Questions worth asking at this point are:

are the aims of the branches of human geography as clear as the aims of climatology and hydrology? Do we want to get past the stages of description, classification, and the collection of facts for their own sake? Does the structure and use of models suggest any way in which the aims of branches of human geography may be clarified, and the work be speeded up to more intellectually satisfying conclusions?

By now many readers will be impatient with the oversimplification that physical geographers use models and human geographers do not. The main reason for this was to try to convince the sceptical that they are familiar with models, perhaps under other names. The balance will now be restored as far as possible.

In his chapter on models in sociology Pahl (*MG*, ch. 7) states that the aim is to model the social institutional framework within which the economy functions. 'Social information thus gains considerably more significance in terms of such conceptual models as a social structure or a social system' (p. 222). Pahl's contribution helps to emphasise three points. One is that models *are* in fact borrowed from one discipline by another, here from sociology by human geography, whatever the dangers of this practice. Secondly, Pahl uses such words as 'structure' and 'system' to refer to the kind of thing being studied; facts and objects have more significance or importance when they are viewed as elements of a structure or system. Thirdly, there is the implication that a need has been felt for a greater intellectual significance, the need to make some worthwhile sense out of a complex mass of facts. In particular, Pahl shows that some of the models which sociologists use in analysing their own problems would be of great use to geographers concerned with questions of environmental determinism, possibilism and probabilism. The model of social change is something which should not be ignored, but ignorance of relevant work in other disciplines is not unique to geography.

The economic models discussed by Keeble (*MG*, ch. 8) date back to the nineteenth century, with those of Lust, Hildebrand, Bucher and Smoller, and while geographers may be more familiar with such 'early' models as that of Rostow outlining the five stages of economic growth, dating from 1955, the lack of communication between economics and economic geography is notorious. Until very recently economists have ignored most spatial or distance factors in their calculations, and geographers have ignored all but the crudest principles of economics.

Keeble suggests that the economist should study economic systems in order to be able to predict and plan. Perhaps for the present purpose the most valuable part of his essay is his

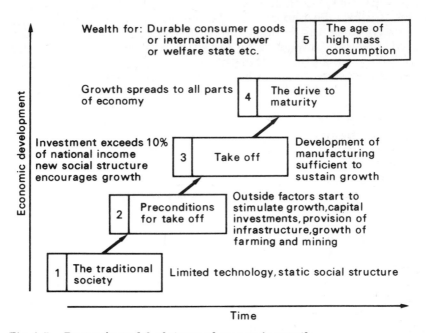

Fig. 1.5   Rostow's model of stages of economic growth

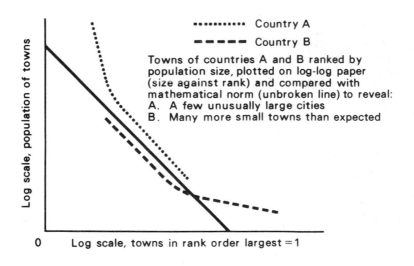

Fig. 1.6   The rank-size model

presentation of three criteria for a model of an economic system. First, the model should single out crucial variables in the system. Second, it should define its concepts precisely enough to permit comparison with the real world; one would add, precisely enough to be tested objectively. Third, it should possess an analytical rather than a descriptive framework. This last point seems to me to be vital. Purely descriptive models will be considered later, but one of the great potential values of the use of models in human geography seems to be the direction of attention to analysis and explanation, to understanding rather than describing.

In a chapter on models of industrial activity Hamilton (*MG*, ch. 10) suggests a much greater variety of activities and aims among economists and economic geographers than in most other branches of the discipline. It is more difficult to see just what the overall aim is, but certain of the points Hamilton makes will be mentioned here and will be taken up in greater detail later.

Again models date back further than recent advertising campaigns would suggest, say to Lannhardt in 1882, and one of the models which Hamilton illustrates is a more sophisticated version of the factors of industrial location which reached school textbooks several years ago. An earlier version of this model simplified the complex decision-making needed to locate a factory until it was believed that the pattern of industrial location observed in real life could be explained as the outcome of seven factors: the price of land, the availability of labour, the availability of capital, the existence of communications, the positions of markets, the positions of raw materials, and the supply of fuel or power. Even in its revised form it may be difficult to recognise this as a model. So many of the models in *Models in Geography* are in graphical and mathematical form, and so much of their explanation prepares one for models to be complicated and unfamiliar, that two or three pages of straightforward English can pass unnoticed as a model.

Of course such men as Weber, Lösch and Isard have taken the model much further, and in particular have attempted to quantify it, until it becomes several chapters in a mathematical language, but the idea is essentially the same. However, it is vital to notice a difference in emphasis between the physical geographers and the economists mentioned here. The former *do* aim to be able to predict, forecast and plan, but only as a secondary function, based on the success of their primary aim of describing, understanding and explaining what actually happens now. In contrast, the more one studies the work of the latter, and criticisms of it, the more one realises that these economists at least are not concerned to explain how industry has located in the past or how it locates now; their primary aim seems to be to work out a model or formula which will allow industries in the future to select automatically the perfect location.

In making certain assumptions at the beginning of the model building process, the economists assume that the industrialist is governed only by purely economic forces. Now while there may be occasions in the future when this will be the case, geographers are well aware that in the vast majority of cases other factors have been in operation. Even to work in the future, the model procedure for 'how to locate your factory' will demand not only that the industrialist considers only economic factors, but also that he has perfect, full and instant knowledge of all the factors operating over the whole of the area in question. No wonder the many equations of Isard are so complicated.

Two points vital to the fuller consideration of models emerge from this. First, in many cases the model builder has to make some assumptions which are just not true: for example, that base level remains constant during a cycle of erosion, that heat from the sun affects the atmosphere only directly under the overhead sun, that farms or settlements have developed on a perfectly flat, perfectly uniform isotropic plain, that man is perfectly rational, perfectly informed, and motivated only by economic considerations. Such fantastic ideas may well be necessary when roughing out the first approximation to the first tentative model, but by the time any worker is testing a model which he really expects to be proved and accepted, it should be sophisticated enough not to need such over simple assumptions.

Second, this emphasis of the economists on how industrialists should locate rather than how they do locate shows that there can be at least two types of model, depending on the aims of the people using them. One type is used to help describe and explain a system as it actually operates in a given part of the world; the other is used to outline the ideal structure and working of a possible system which might be made to operate in the future.

To set against such ideal abstract models as those of Isard, which may well convince the sceptical geographer that models are of no use to him, the reader is urged to study what Hamilton and others in *Models in Geography* have to say about input—output models. Geographers, particularly regional geographers, have long claimed to study interactions on the earth's surface; for example, between man and land, between one economic system and another within a region, and so on. A strong criticism of this traditional type of geography has been that it is neither quantitative nor consistent in its approach. If any geographers are genuinely interested in interactions on the face of the earth, and perhaps doubt whether they are precise and consistent enough, then the input—output model is one which they should consider very seriously as a possible useful tool (*MG*, p. 413).

Wrigley states at the beginning of his contribution (*MG*, ch. 6) that there has been little attempt to construct models in demography. He implies that the aim of demography is to predict population change. Two of the models currently used to help achieve this are the age—sex pyramid and the demographic cycle; both are being used to gain some insight into the concept of the optimum population structure. These two features provide relatively familiar diagrams, and are often mistaken for models in themselves. However, even at the present stage of qualitative hypotheses, these diagrams must be accompanied by a full verbal description of what they represent and a suggested explanation of how the process represented works. Eventually the model must be refined to include precise numerical statements of such things as numbers of people, proportions of male and female, proportions of each age group, rates of increase and so on. Moreover, these must not be simply a collection of numerical descriptive statements, but must be set out as equations representing such things as effect of rates of growth on population size, and effects of population size on rates of growth.

One of the most useful lessons from Wrigley's chapter, however, is only marginal to the study of models. He states that one functional model in demography now suggests that agricultural technology is a function of population density, rather than population density being a function of the level of technology. The argument runs that population densities do not necessarily depend on the food resources of the area, but on the demographic structure of the society.

This hypothesis may be difficult to accept in Europe or North America, but in the study of Africa it may well be the key to explaining great contrasts in population densities when the geographer cannot find similar contrasts in the physical environment nor in the farming sufficient to account for them. Wrigley (*MG*, p. 204) states, 'where fertility and mortality schedules vary considerably, there is much greater scope for regional diversity (in population density)'. Later (p. 213) he writes, 'to neglect demography is to shut oneself off from one of the most important sources of insight into the question of where people live, in what numbers and by what means'.

Notice that this idea that population density affects farming technology is put up as an hypothesis to be tested and possibly

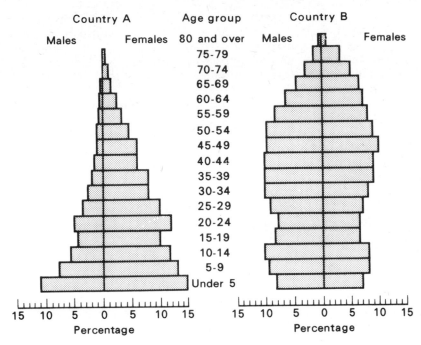

Age-sex pyramids as models of population structure
COUNTRY A Young population with excess of females
COUNTRY B An ageing population with excess of males

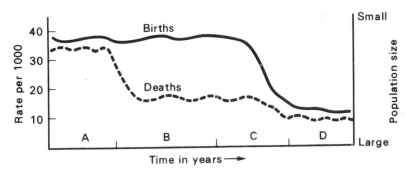

The demographic cycle
A High birth rate, high death rate: stable, small population
B High B.R. decreasing D.R. rapidly growing population
C Decreasing B.R. low D.R.: Decreasing population
D Low B.R. low D.R stable population at a new size

Fig. 1.7   Population models

disproved. To some readers this particular hypothesis may seem so bizarre that at least they can see that it is an hypothesis, and, as yet, nothing more. This helps to point up the perennial danger that many model hypotheses put up for test, seem so acceptable, so plausible, so entirely logical that they are accepted as proven truth without any more ado.

The marginal lesson is that because one has been more excited by the content rather than the structure or the use of the model, preoccupation with models may have some valuable byproducts. It might be safer and more profitable to study what *ideas* other disciplines and branches of geography are developing and pursuing, rather than being completely mesmerised by the shiny new tools they are using in the process.

It would be fascinating to know how many teachers, how many lecturers, how many authors, when dealing with the urban geography of the British Isles have carefully ignored half the towns and cities. When challenged about Inverurie, about Ballycolla, about Caersws and about Alford they may well rationalise that such towns are too small to bother about, there is no information available, there is nothing special about them, or some other unconvincing argument. The truth is that geography has for too long concentrated on the special and the unusual, and for too long has been inconsistent in using a particular method in one situation and not in another.

British geographers for too long concentrated on the industrial towns and the ports, leaving the others alone except for some half-hearted references to county towns and defensive sites. For example, anyone who is not a specialist in Irish affairs might well believe that Derry, Belfast, Dublin, Cork and Limerick, all ports, are the only towns. Whether conciously or not, the descriptions of the industrial cities and ports in Britain have relied not only on the fact that these places are important and well documented, but also on the fact that economic models are in existence to make it possible to explain their sites, location and distribution, rather than to describe each one individually in a kind of catalogue.

While perhaps not every town of 5 000 inhabitants deserves individual attention, the sheer number of small towns distributed throughout Britain does. They are features of the geography which in aggregate cannot be ignored, and moreover, elements of a system for which there has been at least an hypothetical explanation since Christaller published his model of central places in 1933. Garner, (*MG*, ch. 9) claims that settlement geography has adopted a new paradigm in that formerly it aimed to study the relationships of man and the land, namely settlements and their particular sites, whereas now it aims to study interrelationships between groups of people, that is among large numbers of settlements. It is true that much settlement geography is now concerned with the distributions and locations of villages, towns, cities and functional zones of cities in space rather than their particular sites at springs or on volcanic plugs, but is it new? Garner quotes the work of Hurd, 1911; Burgess, 1925; Christaller, 1933; and Hoyt, 1939. What is new is that some geographers and geography teachers are beginning to adopt some old models put forward by these sociologists, models which are already being superseded by the models of Lösch, Berry and others.

While one may be rather embarrassed that geographers are adopting as new and valuable tools forty-year-old models already discarded as obsolete by sociologists, at least it is encouraging that they feel the need not only for models, but also for a new approach to their work, a new paradigm. The new approach is the important thing, not the particular model used or even the use of models at all. For example, one would no longer be content with describing the individual sites of all the small towns in Lincolnshire, listing the sizes of population and the types of industry, and then giving the inevitable *historical* explanation of how each town came to be as it is. The need felt now is for a description dealing with all the towns at once, and a *geographical* explanation that is one which explains why they are distributed as they are, and how they are related in space to each other and their surroundings.

For the old type of approach one could visit each site, or study it on a map; one could obtain the population and industrial facts from the census; one could piece together the narrative from the historical records. The work was a matter of historical research and plodding compilation. No model was necessary. For the new approach the answers are not to be found in the sites, in the latest statistical publications nor, in the local libraries. An explanation for the dimly perceived functioning of the system has to be imagined. An explanation has to be imagined, then it has to be formulated as a clearly structured model which can be tested against reality to judge whether it is a logical and true explanation of that reality.

For the example in question (with apologies to Christaller) one may suggest that the following is a model of the urban system of Lincolnshire. The towns are regularly spaced so that each is positioned to serve an area of agricultural land with shops, professional services, schools, hospitals and entertainment; that the towns are distributed in a triangular pattern, serving hexagonal areas; and that the services required less frequently are fewer in number, located in a small number of large towns, spaced further apart from the rest. The assumption is that the majority of settlements in Lincolnshire are central places, evenly distributed throughout the area to serve the population.

The complete model would need a fuller verbal description and explanation, a diagrammatic map showing the triangular layout, and precise quantification stating how many towns, how far apart, with what sizes of population, number of services and so on. Those not impressed by central place theory will be able to marshal many objections to this. What about Grimsby, the resorts, Scunthorpe? Christaller specifically excluded break of bulk points and resource based towns; he was not aiming to explain their locations. What about the Wolds and the Fens which distort those sinister hexagons? This vexing problem will be examined thoroughly in chapter 5.

But Lincoln is a defensive site, a gap town, a cathedral city, a regional centre, a county town, a service centre, an engineering town, a route focus, a bridging point. Perhaps the sceptic can add to this list, and perhaps he can ask himself, *as a geographer*, just what does he really aim to understand about Lincoln, the settlements of Lincolnshire, and the settlements of any other part of the world?

We *suggested* this model of the urban structure, one of many possible explanations which can be tested against the facts to see whether it is valid. 'Please, sir, how do you spell "irrelevant"?' 'Look it up in the dictionary, idle boy.' 'How can I look it up if I don't know how to spell it?' In terminology which might be over the head of the boy who thinks he has got 'sir' over a barrel, set up an hypothesis and test it. I think it is spelt IRELEVANT. Test against the book of words; try again. Set up my second model, IRRELAVANT. Bad luck! try again. Third model coming up, IRRELAVENT. How can I be so near the truth and yet be wrong again? Model mark IV ready for testing, IRRELEVANT. Match up with the dictionary; how clever I am!

The work of testing the model against the facts of Lincolnshire would obviously be longer, more varied, much more open to complete failure, but basically involve the same process. One can measure whether the towns are regularly spaced in a triangular pattern — having decided how much latitude one can tolerate. One can tabulate the services of every town, and test whether the numbers correlate with the sizes of population. By means of questionnaires and fieldwork one can verify whether in fact the small towns do serve small areas and the large towns large areas. Perhaps the most dissatisfying result of all this is that one will neither have proved the model true, nor will one have investigated every aspect of the urban geography. The model will merely stand as an acceptable theory until a better model comes along. During the course of the study one will come across so many other topics, problems and questions worthy of study, that one will be convinced just how simple any one model is, and just how tiny a fraction of reality it represents.

The situation in agricultural geography outlined by Henshall (*MG*, ch. 11) reminds one of the situation in geomorphology in that there appears to be neither one basic aim in the discipline, nor one generally accepted model. There seem to be four different types of explanation in the study of the distribution of farming at the moment.

First is the explanation by reference to the relief, climate and soils in different parts of the world. We are so thoroughly familiar with this approach that we do not consciously realise that at some stage in the past, someone has said, in effect: 'The farming varies from place to place; I suggest that it varies because the relief, climate and soil vary from place to place.' Again, this seemed such a plausible hypothesis that it was obviously accepted as so near the truth as not to need testing and proving. Most geographers now have reacted strongly against physical environmental determinism but many workers are still convinced that physical conditions are the overriding factors.

Second is the explanation by reference to economic factors. Here we have the other extreme, embodied in another very familiar model, the concentric zoning of agriculture round the isolated city, first put forward by Von Thünen in 1826. The isolated city is in the middle of the flat plain, over which the climate and soils are uniform from end to end. Von Thünen suggested that farming varied with distance from the city, according to the cost of production, the cost of transport to the market in the city, and the price paid for each commodity in that market. This is no more a complete view of the real nature of things than is the physical model. In the physical model, economic forces are ignored, in the economic model physical factors are omitted as one of the basic assumptions. While Von Thünen's model is no longer as universally applicable as Christaller's model, Chisholm (1970) gives a convincing account of its use at local and international levels.

Third is the explanation by reference to the spread of ideas. The diffusion model hypothesises that farming varies from place to place because ideas and techniques originate at specific points and then spread outwards among the farming communities. Moreover the model includes a mechanism which simulates a way in which this diffusion could take place.

Fourth is the explanation by reference to farmers' decision-making processes. The hypothesis here is that the farming in an area depends on the farmer's decision when he has taken into account what partial knowledge he has of his soils, the likely changes of weather, and the state of the market, all coloured by his cautious or bold nature. Again the model includes a representation of the decision-making process in the form of a game.

From these outlines, two points are clear. Hypotheses two, three and four are so unusual that we would demand that they be tested, unlike hypothesis one. In addition, we realise that each contains only part of the truth, and some future satisfactory model will have to combine the essentials of all four, possibly with other factors not taken seriously as yet.

The changes taking place in geography in the third quarter of the twentieth century are usually bracketed together as 'the Quantitative Revolution'. In some books and journals one finds lists where correlation tests, sampling techniques and models are lumped together as though they were things of the same order. It has been implied already that fieldwork, sampling, and statistical and mathematical operations are techniques employed in the process of testing a model. The model, in fact, is a much larger, much more important, much more all-embracing tool than the quantitative techniques. However, one would hesitate to write of a model revolution; the main revolution in geography seems to be the demand for a discipline which concerns itself with structured concepts rather than with narrative and description. The real development, in which models are essential tools, is the revolution to properly organised and directed research.

Geographers are finding challenging intellectual work, not in dealing like the fire brigade with a flood of uncalled-for data, but in stating their hypotheses in a way which can be tested, and then testing them ruthlessly. Many theories and principles of geography in the past were never called models, but models in fact they were. The difference now is that hypotheses are not acceptable as principle and theory until they have been rigorously examined. So the objection to the old qualitative geography is not that it neglected quantitative data, but rather that it did not state its suggestions in a manner sufficiently precise for anyone to test their validity.

Many of the models considered briefly in this chapter have made provision for prediction. The function of the human geographer is neither to predict, nor to tell people what they ought to do. The work which distinguishes the geographer from the geomorphologist, meteorologist, sociologist, demographer, economist and others mentioned is the study of the interrelationships of physical and economic phenomena on the surface of the earth. If there is a need for work more intellectually satisfying than in the past, then the study of these systems on the earth's surface offers a proper course of action for the geographer, and the use of models offers the most effective means that we know at the moment.

The main purposes of this introduction have been to illustrate the nature of the beast called 'model', so that the word carries some meaning in the later discussion, and to suggest that the changes in geography have created a conscious need for the type of work in which the use of models is appropriate. The following chapters will make a critical examination of aspects of models, and will not seek unnecessarily to advocate their use.

2
Definitions and types of models

It soon becomes clear that there are more types of models and a greater variety of uses than is implied in chapter 1, but just how many types is not so readily apparent. The problem of definition is aggravated by the fact that various writers, each with his own particular approach, have produced different definitions without comparative assessment. For example, McCarty and Lindberg and Chisholm (1966) have their own clear, but specialised, ideas of the model; while Cole and King (1968) and Chorley and Haggett (*MG*, 1967) give lists of definitions with little analysis. The list presented below has been compiled from the works listed in the Bibliography (p. 155). These definitions are not credited to authors, partly because many of them are used by several authors, partly because the main purpose of this list is as a starting point to sort out some worthwhile ideas about models.

A model has been defined as:

A
    1. An hypothesis
    2. A theoruncula or theorita
B
    3. A theory
    4. A law
    5. An explanation
    6. A rule
    7. A theory of the structure of something
    8. A formalised theory
    9. A general principle
C
    10. A description of phenomena in mathematical terms
    11. A vehicle to carry information
    12. A description
    13. A miniature representation of a thing
    14. A representation
    15. An abstraction

16. A picture of how a system works
17. An abstraction of a system
18. A demonstration

D

19. An equation
20. The scientific method
21. A way of thinking
22. A proposed method of research
23. A structured idea
24. Something on which to work
25. A mental construct
26. A way of looking at things
27. Reasoning about the real world

E

28. A frame of reference
29. A framework in respect to which a subject is described
30. An organisational frame
31. An ideal

F

32. A synthesis
33. An analogue
34. A set of constraints
35. A simplified version of reality
36. A psychological crutch

Some immediate consideration will be given to most of the points raised by this list, but all the definitions listed here are not of equal importance, and it is emphasised that some of the implications will not have been followed through until the end of the book.

Sections A and B draw attention to the commonest problem encountered by anyone trying to understand what is meant by the word model. Strictly speaking, an hypothesis and a theory are the same thing, a supposition put forward to account for something not understood, while a law, an explanation and a principle are things applied to a phenomenon which is understood. As will be demonstrated in detail later, the 'thing' used as a model for a phenomenon can be used both in the early stages as an hypothesis about the phenomenon, and, later, if proved, as an explanation of that phenomenon. Thus sections A and B of the list are not contradictory, they are merely parts of the truth. The problem in reading about models and reading substantive works which use models is that the authors tend to apply these words and phrases imprecisely. In particular, Harvey (1969), who deals with models most exhaustively and generally is concerned to be very precise, seems to distinguish the word

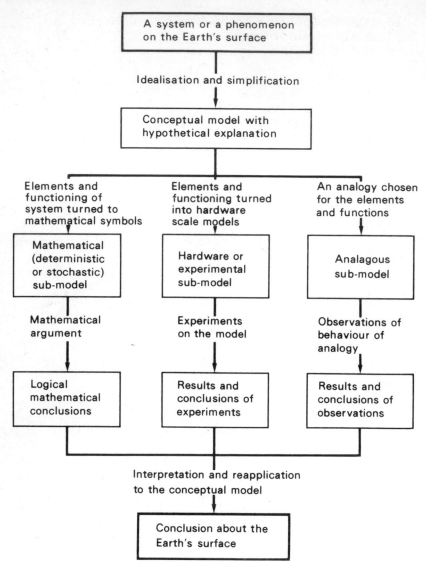

Fig. 2.1   The use of models, after Chorley. Note that, for the geographer, the process begins and ends with phenomena on the earth's surface

theory from the word hypothesis, using the word theory to mean tested and approved explanation.

Item 13 reminds one that geography is only one of many fields of research in which models are used, and that the word may mean different things to different geographers. When a map is used as a model by a geographer, it is a miniature representation; but when a contraption of billiard balls is used by a physicist as a model of an atom it is a gigantic representation. Thus while a particular definition may seem appropriate in one context it may be ridiculous in another; a more comprehensive definition is needed. Within part C of the list we have the statements that a model is a description, a representation and an abstraction of some phenomenon in the real world. Certainly the model, even at the hypothetical stage, must contain some description of the phenomenon in question, some representation, whether in words, drawings or figures, and inevitably, because it is only describing and representing one aspect of reality, it must in some sense be an abstraction. Again the definitions contain parts of the truth.

When a model is defined as 'a description of phenomena in mathematical terms' or as 'an equation' this raises a matter much more important than the fact that some workers, particularly economic geographers, have given only a part definition in their own terms. One of the definitions which Harvey (1969) reveals so clearly is that the word model to some people means a standard procedure in mathematics, logic, or experimental work. Items 20, 21, 22, 26 and 27 make sense in the light of Harvey's exposition of this matter, and George's similar points (*MG*, p. 43).

One begins to fear that the simple word 'model' is burdened with too many meanings when evidence from other works assures one that item 24, 'something on which to work' is also a correct, part definition of the truth. Maps, mathematical equations, diagrams, and contraptions of wood, metal, plastic and clay can be both models in accordance with the other definitions already put forward, and also devices on which mathematical, graphical and physical tests, operations, or experiments can be carried out. Part of that collection of words, diagrams and equations which comprises the model may, at times, form the 'apparatus' on which experiments are carried out, to test the hypothesis of that same model.

Surprisingly, four of the thirty-six definitions refer to one secondary use of some models. An ideal situation may well be hypothesised before research begins, not with any naive idea that the real world is ideal, but rather to see how close reality approaches the ideal. However, the framework referred to in items 28, 29 and 30 is more likely to be an end product of successful research, available for use in teaching or further investigation under different circumstances.

Section E is included so that the list may be as complete as possible, but the points here vary widely in their importance. The model as a synthesis will be discussed in chapter 4 under the name 'compendium'. To call a model an analog, raises one of the most difficult problems in the definition and use of models, and this also will be dealt with in detail later. But it is worth emphasising here, that whatever else the model is, it is a simplification of reality, both of necessity as soon as we symbolise reality in words, pictures and figures, and of intent, because the first hypothesis attempts to offer the simplest possible explanation. This aspect of the simplicity of models is necessary and acceptable. Where models are criticised for their overcrude simplicity, is not in their symbolism, not in their first, tentative explanation, but in some of the constraints they lay down. If an hypothetical explanation of the distribution of towns or farming will work only on an isotropic surface, then some critics would reject such a simple model out of hand, and demand a more elaborate one which can explain such phenomena in the scarplands and in the mountains.

Certain definitions from each section have been purposely ignored until now. They are the ones which refer to structures, systems and mental constructs. It will become clear as this analysis proceeds, that some models are descriptions of static phenomena, and nothing more. In this sense, the word 'model' is no more than new jargon for the kind of generalisation geographers have been making for centuries. However, many models are well structured hypotheses about the arrangement and functioning of systems on the earth's surface. It has already been argued in chapter 1 that while such models are not new in physical geography, in human geography they have not often been consciously formulated, and that the conscious use of such a tool can result in the understanding of the geography of the earth's surface, rather than a general catalogue of its features.

At this stage in the analysis, it is suggested that there are either several distinct types of model, or that models have several distinct types of use. Whether there are different devices with the same name or different functions of the same type of device must remain an open question at the moment. In geography the word 'model' seems to be applied to six devices or uses:

1. as an hypothesis about the nature and operation of a system on the earth's surface;
2. as an accepted demonstration and explanation of the operation of a system;
3. as an apparatus or test bed on which to test the hypothesis or demonstrate the laws, when such things cannot be done in the field;
4. as a standard procedure for carrying out a complete piece of

research, the actual test of an hypothesis, a statistical test or
a mathematical calculation;

5. as a norm or ideal with which to compare reality;
6. as an analog or analogy for the real thing.

Even if it is not the same device which is used in six, *or more*,
different ways, the several devices seem to share a common
structure. They contain a description of the phenomena in
question, an explanation of the functional relationship of the
phenomena, and a precise representation in maths, graphics or
hardware on which certain test or demonstration procedures
may be carried out. The components vary in relative importance
according to the particular use of the model.

A word of warning is necessary here. The outline of the
model suggested above is in my view the most important. But it
will be shown that there are other models, especially those
borrowed from other disciplines, which do not necessarily
include a theoretical explanation of the system which they
represent. Again, some writers insist on calling these devices
analogies rather than models, but the word analogy has many
other meanings.

To call a model 'a psychological crutch' may seem to be derogatory, but this sums up well the fact that models have
developed out of the need for some device to help the mind
comprehend reality. Rutherford, Logan and Missen (1968)
elaborate on this point as follows:

> The reasoning processes used in these . . . models is very
> much the reasoning we all use in our daily lives. Rather than
> consider simultaneously an unmanageable host of variables,
> we abstract from reality, and, in a discipline, we erect what
> amount to hypothetical models which deal with only part of
> a selected situation; generally we deal with variables one at a
> time. In trying to explain patterns that we observe, we follow
> the rule common to all science: we select the simplest
> explanation first and only move on to a more difficult
> explanation if the situation warrants it.

For those who are completely satisfied with the traditional
aims and methods of geography the use of models may appear
to be unnecessary. But in both Harvey (1969) and Chorley and
Haggett (*MG*, 1967), the two most substantial books about
models in geography yet published, two distinct elements are
apparent. Not only do these books expound and advocate the
use of models in geography, but equally, if not so obviously,
their authors assume that readers will adopt a new concept of
the nature of geography, a new aim which makes these methods
necessary.

Chorley and Haggett (*MG*, ch. 1) advocate that geographers should investigate a new type of problem, the spatial arrangement and interaction of phenomena on the earth's surface, rather than such things as the differences between regions, how a landscape came about, or the reaction of man to his physical environment. Grigg (MG, ch. 12) echoes the same theme when he states that, while geography was once concerned with the properties of objects, it is now concerned with their functions and relationships. For the former descriptive cataloguing has often been considered sufficient, but for the latter a functional explanation is necessary. For this new aim, models are much more important, and when Rutherford, Logan and Missen (1968) state that 'the value of a model . . . is to be judged in terms of its explanatory value', the emphasis is clearly on the word 'explanatory'.

One of the most disappointing aspects of Harvey's book *Explanation in Geography* is that he does not provide what one can feel with any certainty is a geographical as distinct from, say, an historical or scientific explanation. At the end, one realises that his complex argument in fact advocates that geography should adopt both the methods and the aims of natural science. This appears to be tantamount to a denial that geography has any identity as a separate discipline. For those still mystified as to why Harvey, Chorley, Haggett and others urge them to use models, the point may be clearer when the reason is seen to be that these writers are more interested in geography as the study of spatial relationships than in geography as the study of regions or landscapes.

Chorley and Haggett (*MG*, ch. 7) parallel the reasoning of Rutherford, Logan and Missen quoted above. They argue that we comprehend aspects of the real world through models; that the complex real world is symbolised in man's mind as a set of simple systems. We end up with models which are simplified structurings of reality and which present significant elements and relationships in a generalised form. As to geography, they suggest that not enough structuring has been done for the subject to remain a satisfying discipline. Therefore they, and others, either advocate a new paradigm or assert that geography has adopted a new paradigm. By paradigm they mean an agreed set of procedures to investigate the problems and work towards those aims. Much mystery about, and objection to, models has arisen because many people have not realised that they are virtually being urged to change to a new discipline (at least to a different aim of geography) as well as to adopt a new method of research.

Chorley and Haggett claim that they want to move from a

classificatory paradigm to a model-based, i.e., explanatory, paradigm. One can agree with this, and approve of their efforts to find models within geography, rather than merge geography with science, without also having to agree with them that their type of geography is either the only type worth consideration or the only type which needs the use of models. For example, one aim of geography, dismissed by Chorley and Haggett, has been the investigation of the influence of the physical environment on man's economic activities. It would seem to be perfectly possible to hypothesise a particular aspect, of, say, the effect of mountains on farming systems, and to test that model, just as one would test the models of Von Thünen or Christaller.

It seems to be equally possible to use models for the study of landforms, soils, farming and manufacturing as for the study of points, networks, patches and surfaces, distributions, diffusions and relative location which interest them. While a change from a study of landscapes to a study of the aspects listed here may necessitate a change to models, it is seriously to be doubted whether the adoption of models as tools by any geographer necessitates a change in his basic interests and aims.

The definitions of models examined in the first part of this chapter are attempts by people who use models to define them in a word or a phrase. As mentioned above, these definitions have not been set out in any logical order, nor analysed in any detail. Their value lies in their origin, in that people who use models refer to them in these words.

A much more orderly attempt has been made to produce typologies of models, but in some respects the value of this work is less clear. Certainly a division into types may lead to a classification, which in turn may help people to understand models more readily, but it will be seen that this division has not been entirely successful so far. The most surprising feature of all the main typologies of models in geographical literature in this country is that they have followed a rather sketchy original, which has prevented a more objective approach.

Ackoff, Gupta and Minas (1962) turned to the dictionary for help. Not surprisingly, they found that the word model can be used as a noun to mean a representation, as an adjective to mean ideal, and as a verb to mean to demonstrate. What they do not seem to have appreciated is that their dictionary was written before that simple word became burdened with more meanings. If only their colleagues, needing a word to label the devices they used, had also turned to the dictionary, or, better still, a thesaurus: then, instead of the word 'model' having to carry at least six new meanings, as outlined earlier in this chapter, these device-users might have found six different words with which to label six different things or operations. In the event, they did not, and Ackoff searched in vain for meanings of which his lexicographer was not aware.

So, for what it is worth, we are told that 'in model building we create an idealised representation of reality in order to demonstrate certain of its properties' (Ackoff *et al.*, 1962). If I splash some green paint on a piece of paper and say 'this is my lawn', I have done exactly the same thing, but I would not claim this to be a model of the type under investigation in this book. Ackoff *et al.*, then suggested the initial breakdown into three types of model, *iconic*, *analog* and *symbolic*, and this has since been followed, with minor changes and additions.

The most recent repetition of this division of models is by the tutors in geography in the Open University. In one set of course notes it forms the only general explanation of models before a particular example is used in some detail (Open University, pp. 1—2). The iconic model is described as a scale model, the analogue model as an analogy, and the symbolic model as a mathematical equation. There is nothing very startling, very difficult, or very useful in all this. The only thing one learns is that models can take different forms, such as a plaster model of relief, treacle flowing down a groove in wood as an analogy for a valley glacier, or a number to represent the density of the road network in an area.

Yeates (1968) and Haggett (1965) have both followed the lines laid down by Ackoff *et al.*, changing only the spelling to analogue, and the examples they offer in explanation, thus:

Iconic model. A three-dimensional model, on a different scale from the original (in geography this usually means smaller), but not necessarily made of different materials.

Analogue model. Properties and materials in real life represented by different properties and materials in the model; this is not necessarily a three-dimensional model, and in fact may be a verbal account of an analogy for the real phenomenon.

Symbolic model. All the authorities mentioned so far refer specifically to mathematical symbols. Some refer to map symbols, but the implication is that words are not symbols for the purposes of this type of model.

From this we derive the useful points that the form which models take can vary enormously, and that there is some idea of increasing abstraction from faithful reproduction, through analogy, to completely abstract symbolic representation. These points should be remembered. On the other hand, their usefulness should not obscure the fact that the simple classification set out above can cause dangerous confusion. For example, there is the set of illustrations given in an Open University paper. The illustration of an iconic model is a scale map; the illustration of an analogue model is a map with different types of roads shown by different widths and colours; the illustration of the symbolic model is a topological map. The fact that it is difficult to find illustrations of these three types, and that an example chosen to illustrate one type also proves useful in illustrating another, strongly suggest that this classification is of little further use.

Haggett (1965) adds three more types of model to the three already listed, but it cannot be emphasised too strongly that these are classified on a different basis. Haggett distinguishes three types of models according to the procedure followed in making use of them.

Mathematical Model. One which involves mathematical calculations which either represent some process in real life or are used to test some hypothesis about real life.

Experimental model. One which involves some practical procedures, exemplified by the chemist or physicist working with apparatus in the laboratory, again to represent a real situation or to test ideas about a real situation.

Natural model. One which involves the procedure of reasoning by analogy, comparing the known with the unknown in the mind or on paper, verbally or possibly graphically.

Haggett attributes these ideas to Chorley, and elsewhere Chorley (in Berry and Marble, 1968) has gone into more detail about these models. It appears that they are meant to be models within which certain procedures are dominant, as distinct from the models of standard procedures which concern Harvey and George. So Chorley subdivides the three models as follows:

Mathematical models
A. *Deterministic.* These models can represent processes in real life where there is certainty that a given cause will result in a given effect.
B. *Stochastic.* These models must be used when there is some doubt about the exact effect of a given cause. In particular, when there is only a probability that a given result will happen.

Experimental models
A. *Scale models* which use the same materials as the original, for example water carrying silt in an hydraulic tank, as a model of a river.

B. *Analog models* which use different materials or even a different system to represent a real-life system, for example kaolin as an analog for glacier ice in a model, or electric circuits as an analog for traffic systems.

Natural models
A. Those which compare the phenomenon being studied with something better known.
B. Those which compare the phenomenon being studied with something which is more accessible for study or experiment.

Even with this subdivision made by Chorley, the models still stress the mathematical, practical or logical procedures which either simulate a real phenomenon or enable investigations into it to be carried out.

Cole and King (1968) and King alone (in Balchin, 1970) have put forward classifications which combine parts of both the iconic—analogue—symbolic set and the mathematical—experimental—natural set. At least in the labels used, they seem to have mixed together a set which classifies the form of models, and a set which classifies the procedures or operations in models. However, it has been shown that one verbal label has been forced to carry more than one meaning, so their classification may be more logical than this implies.

Cole and King (1968) take scale models, simulation models and conceptual models as their three main types, and subdivide each type.

Scale models
A. *Static*, e.g., a relief model
B. *Working*, e.g., an hydraulic tank
C. *Two-dimensional*, e.g., a map

Simulation models
A. *Stochastic, probabilistic*
  1. to simulate physical processes which are believed to be random;
  2. to simulate human systems which seem to follow the laws of probability.
B. *Mathematical, deterministic*
  1. to test physical processes;
  2. to provide a norm with which human behaviour can be compared.

Conceptual or theoretical models
A. Those which provide the paradigm for geography.
B. Those which help one to formulate types of explanation.

At first sight there appears to be a mixture of the form of

models and the function of models in this classification. But whatever was intended, and this is not explicit, this classification can in fact be used to stress the functions of models. Some tidying-up is necessary first. Either 'scale models' should be subdivided into two-dimensional models and three-dimensional models, or two-dimensional models such as the map should be included with static models. Then one can see that scale models can be divided by their function, either as static representations or as dynamic working models on which experiments can be carried out.

In a similar way the functions of the stochastic and deterministic models stand out, in that they can simulate actions and processes in physical and human geography. Likewise, the conceptual or theoretical models can function to outline the aims of geography and to provide examples of the type of explanation geographers are seeking to give.

One may be reading more or less into the classification set up by Cole and King, but in their *Quantitative Geography* it is not clear why they have put forward this particular set of types of model. The chapter on models is a collection of facts about models without analysis. King (1970) later put forward a modified version of this classification.

Scale models
Maps
Simulation and stochastic models
Mathematical models
Analogue models
Theoretical models

Here maps appear in their own right and analogue models appear as a new type. This may result from the very tricky problem of exactly where to assign maps and analogues, but the same indecision is seen in that the classification is not by form, procedure, function or any other explicit attribute.

Anyone attempting such a classification may be in danger of trying to sort out widely differing things to which the name model has been applied indiscriminately. Unless the name has been applied on some systematic basis, and unless there is one concept or set of concepts behind the name model, then it may be necessary to ignore the name and examine the phenomena. This will involve the strong possibility of having to reject some things at present called model, and of having to recommend new names for closely connected phenomena.

In mentioning that some commentators can name the map as the best example of three supposedly different types of models, or that an example of a model for one writer is an example of an analogue for another, it becomes clear that certain pertinent questions must be asked about these terms and classifications. The questions will not be rhetorical, because the answers are not obvious, and perhaps different readers will answer them in different ways according to their varying experience.

The basic point about the iconic model is that it has the appearance of the real phenomenon. Thus at once we imagine that it will be much easier to make an iconic model in geomorphology than in population geography, because one can reconstruct the appearance of, say, a coastline much more easily than one can reconstruct the 'appearance' of the distribution of population. The key question here is, given that the model must have the appearance of the real thing, must it also have the function? Must the model of the cliff coastline also have some mechanism to reproduce miniature waves so that the model will function like the real thing, and erode the cliffs here and deposit debris there? There are many such working models in departments of geomorphology and in hydraulics research stations, but the question remains, does an iconic model, by definition, represent just appearance, or both appearance and function?

The same question may be asked about the analogue model; is it an analogy of just the appearance and structure of the real phenomenon, or must it also contain an analogy of the functions and operation? If the iconic model looks like the real thing, then the analogue model has some basic similarities with the real thing. One of Chorley's examples of an analogue model (in Berry and Marble, 1970, p. 49) is that of kaolin mud flowing down a model valley. This represents the operation of the real thing much more than the appearance. King (in Balchin, 1970) compares the circulation of material in the landscape with the circulation of blood in the body, and in this example no one would expect the appearance to be the same.

By the time the classifiers reach stage three they have already shifted their ground. The very term 'iconic model' tells one at once that the appearance is being modelled, and the question is, in what materials or medium? But the term 'symbolic model' gives one exactly the opposite type of information, making it clear that symbols are the medium in use, but leaving the question, what feature of the original phenomenon is being modelled? For the moment this particular question will be answered, in order to ask a more important one. Given that the appearance or form of the original phenomenon is being modelled or represented by symbols, then what kind of symbols do Chorley, Haggett, King and the rest have in mind? Most of the time the answer seems to be mathematical symbols, but others are mentioned, and the word symbol has a much wider meaning than this.

The rest of the terminology is equally ambiguous. It may be clear enough that the mathematical model is in the form of mathematical symbols, but, as suggested above, it takes time and familiarity with several pieces of work for one to realise that the mathematical model represents some aspect of the function or operation of the original phenomenon. However, while the experimental model is also concerned with the operation and functioning of the original phenomenon, there are two inconsistencies here. The emphasis, more often than not, in the references made by writers to experimental models, is that these are used not simply to represent the functioning of the phenomenon, but to test possible changes which could be made in the functioning of the real thing. For example, the impression given is that a set of mathematical equations may attempt to represent the functioning of a depression, but that a working model of a coastline is used to try out, in turn, the effects of groynes, sea walls, dock construction and so on.

Now this is only an impression, for the two 'types' of model are not mutually exclusive. By changing values in the mathematical equations, 'experiments' can certainly be carried out on so-called mathematical models. Similarly, the so-called experimental models also represent the functioning of the real thing. In fact, for any of the experiments relating to possible changes in the system to have any value, then the model must first represent the present functioning very faithfully indeed. Therefore it is suggested that both these models are, simultaneously, models of the functioning of systems rather than of their appearance, and models on which experiments can be carried out. The alternative name of 'hardware models' for experimental models is no better, because hardware models have other uses too.

Finally, the term natural model at the end of the list, really refers to all the other possibilities left over when those in mathematical or hardware form have been taken out. Some workers probably take natural model to mean a natural analogy which can represent the function of the system being studied and on which it is easier to carry out experiments than on the real thing. Chorley (in Berry and Marble, 1968, p. 49), for example, being unable to observe the formation of drumlins, claims to have gained some insight into the forces producing them and governing their shape by studying the production and formation of birds' eggs.

In this example we have some claim that the operation of one natural phenomenon can represent the operation of another, but it is difficult to see how it would be easier to carry out experiments on one real-life system than on any other. There-fore, for this part of the classification to make more sense, and for it to be complete, it is suggested that the phrase 'natural models' should be replaced by the phrase 'other models' and then that this should include all other methods of both repre-senting and experimenting on the operations of systems. There would be a greater variety of form than is implied at the moment in that the operations could be represented by diagrams and manufactured analogies as well as by natural analogies. Experiments as such would not be appropriate to this group of models, but the parallel with other models is that, while the mathematical models demand emphasis on mathe-matical procedures, and the experimental models on practical procedures connected with the hardware, these demand emphasis on verbal and logical procedures.

Iconic, analogue and symbolic models, then, represent the appearance and form of the phenomena being studied. Above all, it should be stressed that these are methods for representing the structure of the systems being studied. Whether by making a hardware model which looks like the real thing, or finding an analogy which has structural similarities, or by putting symbols on paper, we idealise and simplify the elements of the system and their arrangement. There may be much simplification, much isolation from reality, but the purpose of using these models is to enable the student to get a clear mental concept of the relationship of the elements within the system.

Mathematical, hardware and natural models, to use current jargon, at one level represent the operation of the phenomena being studied. These models contain mechanisms which can represent the functioning of the systems in question. Whether by using equations which represent the flow of energy and matter through a system, or by making a model which works like the real thing, or by finding an analogy which works in a similar way, we try to get an idea of how the system works.

The purpose of using models of any kind is to provide a clear if oversimplified idea of the interaction between the elements of a system. Mathematical, hardware and natural models have a further application at another level. It has been made clear that they are used for experiments to test the kind of changes that man would be able to make in real life; but, and this is equally important, many experiments, trial runs, or tests have to be carried out simply to get the model to operate like the real thing in the first place. To enlarge on this point here would be to lose the thread of the particular argument of this chapter, but it must be stressed that this testing of the model against reality is one of its major functions in helping us to understand reality. This aspect is so important that it will be taken up and examined in detail in later chapters.

Two examples may be helpful at this stage to illustrate in specific terms the generalisations about form and function of systems which have been used several times in the last few paragraphs. A soil profile can be presented either as an iconic model or as a symbolic model. In the former, different coloured materials such as plastic beads, gravel, sand or sawdust could be piled on top of each other in a glass tube or case to represent the appearance of the A, B and C horizons; in the latter, a coloured diagram could be drawn to represent the horizons. In both types, either a structure has been made, or symbols have been drawn, to represent what are considered to be the elements of the system. In this example the system is the complete soil with air, water, plants and animals moving through the mineral debris, and the elements of this particular model of the soil are the three levels, A at the surface containing roots, B lower down with a different chemical composition, and C the unweathered parent material at the base.

This example, so far, demonstrates another essential feature of a model, the simplification, idealisation and abstraction. To be able to see three distinct horizons in an exposed soil is often an act of faith, so these are really an idealisation and simplification of the real thing. Further, it was stated above that in this model the horizons are considered to be the elements of the system. In a different model, different sizes of mineral debris, or the minerals, air, water, plants and animals could be considered as the elements. Thus one particular model is an abstraction of those elements which concern a particular student.

Having made an iconic or symbolic model of the appearance of the elements of the soil system, an experimental or natural model could then be made, first to represent the working of the system, then to carry out experiments on it. In the case of a podzol, the iconic model could then have water draining down through it to represent the leaching of salts from the A horizon into the B horizon. The symbolic model would have to hypothesise as to which salts are dissolved, how they are moved downwards, and how they are redeposited at the lower level. An advanced model might require quantitative statements of amounts of percolation, rates of leaching, proportions of soluble matter moved, and so on. In this way the working of the system has to be represented either by a working model or by a collection of diagrams, verbal statements, verbal reasoning and mathematical equations. To produce an accurate natural model of this kind may involve much trial and error, comparing with, and testing against, reality, over many years.

The same devices may also be used for experiment in the sense of trying out procedures on the model before they are tried out in real life. The effects of drought, irrigation, deep ploughing, different types of plant cover, the addition of chemical fertilisers or organic manure can each be tried out separately, and in turn. In this we see another valuable function of the model, for, like all other experimental procedures, it permits control which is not possible in the full scale original, and it permits the change of one factor at a time by a measured amount. Thus the effects of irrigation could be predicted, or a policy of the use of a certain type of fertiliser could be advocated. But it must be stressed that such prediction would be reliable, and such advocacy responsible, *only* after the structure and working of the soil system have been accurately modelled and properly understood in the first place.

Because the word model too easily conjures up visions of papier mâché, polystyrene and sand tables, another example will now be described, to help dispel such visions and to illustrate some other aspects of the concept of the model as developed so far in this argument. The example of a system to be chosen this time is the dairy farming in a particular country. Some readers may already have objected to the word 'systems', and may now not be prepared to apply it to a topic as familiar in human geography as dairy farming. In this context one has to admit that the use of a model does make full sense only if one accepts the new paradigm of geography as well as the new methods. It is still possible to explain how the dairy farming in a particular country came about, or how it has produced a particular landscape without reference to models. But if one accepts that the main concern in geography now is to explain the distribution and interaction of phenomena on the earth's surface, then this can best be done by thinking of dairy farming as a system and constructing a model to represent and explain it.

Once this is accepted, the elements of this particular model are the dairy farms within the particular country. It would be tedious and entirely unnecessary to construct an iconic model of all these farms, say little green patches on an enormous relief model of the country, as was done for the soil profile. Similarly it might be difficult and unnecessary to look for an analogy which would represent the distribution of the elements of this system. Clearly a symbolic model is called for, specifically a map which by means of symbols or shading shows the distribution and location of the farms in question. For the sake of the argument, it will be stated that this map, this symbolic model of the structure of the system, shows that the dairy farms are located in rings around the big cities, and in a distinct zone in the northwest of the country.

It is now necessary to model the forces operating to produce this pattern. The system being modelled is dynamic in the sense that the distribution of dairy farming has changed and will change again in response to changing economic factors. But the purpose of this particular model is to represent the factors in operation now. Hardware models have been used for such a purpose. Hamilton (*MG*, ch. 10) describes how strings, weights and pulleys have been used to demonstrate how several forces operating in different directions result in the location of a given economic activity at a particular place. This may work for one farm or one factory, but not for thousands of farms.

A mathematical model, particularly one which can be run through a computer to deal quickly with the hundreds of thousands of individual forces which have operated on the thousands of farms, may well be useful, but by itself this will only simulate the result, and will not necessarily either represent the processes realistically or suggest an explanation. So a natural model is called for in this instance. In explaining why the factors affecting dairy farming have resulted in the particular pattern seen on the map, it may be profitable to reason by analogy. If another type of farming has been well investigated, modelled, and explained in a satisfactory manner, then parallels may be seen in the present problem. But in the example chosen, it is quite probable that neither hardware models, mathematical models, nor analogies will represent the functioning of the system, and it becomes necessary to try another approach implied in the term 'natural model' as used by Chorley.

One or more hypotheses must be set up and tested against reality, and in the case of farming many hypotheses may be necessary. It may be suggested that the farms have been located where relief, temperatures, rainfall and pasture are suitable for dairy cattle. This might hold up in the northwestern zone, but not round the cities. So it may then be suggested that the farms have been located where they can make the highest profit on milk. This holds up round the cities but not in the northwest. Not only does a combination of these factors operate, but further investigation reveals that other factors such as tradition in the northwest, agrarian momentum round the cities, tariffs, milk marketing schemes, and so on are as important as prices in the city markets, the cost of transport and the costs of production in helping to decide just where dairy farming will be carried on at a given point in time. Moreover, as Chisholm (1966) argues, the problem may have to be approached from two directions. One must consider not only the traditional idea that the farmer asks 'what shall I produce here?', but also of the more modern farmer who says: 'this type of farming is most profitable, now where is the most economic location for it?'

To sum up, the complete model of the dairy farm, if it attempts to represent the structure, to represent the operation of the factors, and to suggest an explanation of these, must be made up of several elements. These elements comprise the map, a verbal description of the factors in operation, a description of the operation of the system in the sense of the farms in all the areas supplying demands from the cities, and, possibly, other farms supplying demands for feedstuffs from the dairy farms. Finally, some analogy with another economic system may help, and a mathematical model might well be included in a sophisticated version of the model to quantify the amounts of milk involved, the volumes of flow in different directions according to the sizes of the markets, cost of transport and so on.

The implication is that a change in any one of the elements or factors will bring about a change in the system and thus a change in the number and distribution of dairy farms. Similarly, experiments might conceivably be carried out on the model to study the effect of any possible changes. Such changes as the growth of population and the increase in demand, the lowering of tariffs, cheaper long distance transport, the rising price of land round the cities could each in turn be hypothesised and their effects on dairy farming evaluated. Again the more accurately the mathematical part of the model represented the operation of the factors in real life, then the more accurately possible changes could be simulated in a computer, and the more reliable predictions could be.

Returning to the point where the argument was left before these two illustrations of models, it is now possible to examine the classification made by Cole and King. At first sight this has as much bias, and as questionable a choice of words, as the other classifications; but Professor King is a leading geomorphologist, and this classification seems to be more relevant to geomorphology and physical geography than to other branches of geography as a whole. I take leave, therefore, to make some changes in the wording which I believe will make this classification relevant to any branch of geography by adding to the meaning rather than changing it out of recognition.

King writes first of scale models, dividing them into static, dynamic and two-dimensional models. By comparison with the other classifications, this is the equivalent of the iconic—analogue—symbolic group of which the main function is to represent the structure of the system. This is really the same thing written in different terms, and is the geomorphologists' part-view of the whole situation.

The simulation models of which King writes, stochastic and deterministic, of human and of physical systems, again are the equivalent of the mathematical—experimental—natural group in their functions of representing how the systems work and of providing models on which to carry out mathematical, statistical and practical tests and experiments. Again King has constructed a classification to deal with all the models of which she as a geomorphologist is aware, but it is not sufficiently general to deal with other models in, say, human geography.

There is no direct comparison with King's conceptual and theoretical models in the other classifications. This attempt at direct comparison raises a question implicit in all work relating to models, but I have not seen it discussed in works relating to models in geography. King's conceptual and theoretical models are models of the aims of geography and of the types of explanation which might be applied to the systems already represented by the other models. Now every other classification stops when it has dealt with the matters of representing structure and function. Never is it written in so many words, but one gets the strongest impression that many workers either never intend to explain, in the sense that Harvey uses the word explain, or, that they consider that to describe the structure and functioning of a system is sufficient explanation in itself. Therefore King's classification, once adapted and reworded to apply to any branch of geography rather than to geomorphology alone, can be most valuable. Her classification forces attention to the two vital facts that one must have a clear aim when using models, and that one should attempt to provide an acceptable and satisfactory explanation of whatever aspect of geography one is studying. For many workers, of course, the hypothetical explanation of the system they are studying, first embodied in the model, is by far the most important element in the work.

There is so much jargon in geography; there are so many examples of one word with ten meanings, or ten words for the same thing, that one hesitates to aggravate the situation. Therefore, instead of a new classification, using new jargon, the elements of the classifications which have just been discussed have been rearranged into a more logical and more explicit classification. The words and phrases will not be redefined. They are taken to have the meanings already discussed in the preceding paragraphs.

I. *Submodels of structure*
    Iconic or *scale*

     Analogue
     Symbolic
II. *Submodels of function*
     Mathematical ⎫   (*a*) *for simulation*
     Hardware   ⎬   (*b*) *for experiment*
     Natural   ⎭
III. *Submodels of explanation* or *theoretical conceptual models*

Three points need to be made to conclude this stage of the argument. First, it is more logical to refer to the things which Haggett, Chorley, King and others call iconic, hardware or conceptual *models* as *submodels* or *part models*. The ways in which models are used in geography now are rarely satisfied by the use of only one of the items listed under I, II and III.

Second, in most cases it would appear that a model must have *three* submodels or part models. It is for this reason that the classification is divided into the three groups of submodels to represent structure, function and explanation.

When a writer refers to a conceptual, hardware or mathematical model, he may mean an idea, a piece of apparatus or a formula, and nothing more. But the way in which several writers refer to these things gives strong grounds for believing that they refer to a complete model of structure, function and explanation, with emphasis on one of the parts. Thus a hardware model has a piece of apparatus as its central feature, which must be accompanied by concepts and calculations; a mathematical model centres on an equation, which must be accompanied by verbal concepts and possibly diagrams. Whether we call the elements of the device models, submodels, or part models, the fact remains that the type of work carried out by means of the device demands representation, calculation and explanation which can rarely be performed by one element. (In dealing with the use of models, it will be necessary to refer to part models. These will be referred to as type I, II, or III as listed here.)

Third, whatever form the model takes — words, figures, diagrams or hardware — and whatever its name, the following characteristics are essential. First, a representation of the structure, the elements of the system. Second, a description of how the system works, how the elements interact. Third, a sufficient and necessary explanation of the system and its functioning. The inclusion of quantitative data and mathematical formulae are essentials to making the model as precise as possible so that it represents reality as closely as possible and can be rigorously tested.

3
Some terminology of models

The types of models described in chapter 2 were selected from a wider range in an attempt to simplify a complex situation and to put the emphasis on a particular function of models. It is now necessary to consider other types, and, in some cases, the same types classified on a different basis.

Chapter 2 stressed the function of submodels in representing the structure, operation and causal factors of systems on the earth's surface. It may be useful, at other times, to classify by the physical nature of the model, by other functions which it performs, by the stage in the proceedings at which the model is used, or by the manner in which one operates the model. Therefore some new names of, or labels for, models arise, and some labels which have been used above reappear in a different context.

1. *The nature of the model*
   hardware, symbolic, graphic, cartographic, photographic, verbal, etc.
2. *Functions of the model*
   descriptive, normative, idealistic
   as an experiment, tool or procedure
3. *Form of the model*
   static or dynamic
4. *Operational purpose of the model*
   to store data
   to classify data
   to experiment on the data
5. *Stage at which the model is used*
   a priori
   concurrent
   a posteriori

This provisional list may prove to be too short or too long as the situation clarifies over a period of time. Certainly section 1 is not meant to be complete, and in the way the words are used by some geographers the terminology may be repetitive. Thus the word symbolic here means mathematically symbolic, which is the most common meaning in connection with models, but diagrams, maps and even photographs are symbolic in another sense. In 5 the word 'concurrent' has been included for the purpose of developing an argument later in this chapter, but it, too, may prove redundant.

Section 1 refers back to the point made at the end of chapter 2. In this context the models are differentiated by their own nature or appearance, and not by any function they perform. The nature of the model *is* worth some consideration to dispel any idea that a model can take the form only of some hardware apparatus or of a mathematical equation. So much of the theory of models in geography has been written by geomorphologists and economists, and so much use of models has been made in those two disciplines, providing the examples, that one might think that model is synonymous with wave tank or computer program.

Harvey (1969, p. 158) mentions graphic models in connection with economic geography but most attention is given to them by McCarty and Lindberg (1966, pp. 55–6). Writing of the graphic space (or spatial) model they describe it as a diagram or hypothetical map of how phenomena are arranged in space. McCarty and Lindberg (1966) combine the mathematical expression of the gravity model proper (see Haggett, 1965, pp. 35–40) with Christaller's diagram of the distribution of settlements and Von Thünen's diagram of the zones of farming round a city, and refer to all of these as gravity models or gravity diagrams. Here we have the clearest example of geographers who use models such as the gravity model so taking it for granted that a diagram is an essential part of the whole device that it is too obvious to mention in so many words.

Any student who carries out research which necessitates the use of models will learn many such things as he proceeds, and possibly never verbalise or formulate certain essential steps or devices. But much misapprehension could be avoided if those who write about models avoided bias, not only to one branch of geography, but also to certain parts of the model. For the novice, it is sometimes important to state the obvious.

McCarty and Lindberg see the graphic model as most useful in the early stages of a piece of research, to help conceptualise the problem. However, a map, graph or diagram may be useful for measurements at a later stage, and for demonstration at the end. For example, the details and implications of Von Thünen's theory of agricultural location, or Christaller's central place theory, would be very difficult to comprehend without the graphic submodels as parts of the whole.

In section 3 the words 'static' and 'dynamic' carry different meanings for different authorities. Chorley and Haggett (*MG*, p. 25) basically agree with George (*MG*, p. 43) that a static model represents the structural elements of a system and a dynamic model represents the functioning of the system. But writing in the next sentence of historical models, Chorley and Haggett reveal that 'static model' could refer to the situation at one precise point in time, while a dynamic model would represent a changing situation over a long period of time. For Cole and King (1968, pp. 465—6) and for Morgan (*MG*, p. 727) the static model, in contrast, is something purely for the purposes of demonstration, and the dynamic model is a much more valuable device on which one can carry out experiments. Perhaps the most valuable lesson to be learned from this mention of the words static and dynamic is of the danger of accepting the first definitions one comes across. Many writers have the habit of setting up labels and jargon as though they were well-known phrases, generally accepted by all other workers in the same field as having an agreed meaning, when they have nothing of the kind. Meanwhile other writers refer to the same words and phrases as having a different meaning or emphasis, yet imply with equal confidence that this meaning is the one generally accepted.

The jargon which refers to models is not so well known, nor in sufficient everyday use, for any of the words to be used without definition and qualification, especially when more than one branch of geography is involved. In particular, the word 'model' itself means different things to climatologists and social geographers, and can mean so many things according to the context in which it is used that in some cases one wonders whether the word model really is appropriate any more. A case in point is item 4 of the list above where methods of storing, classifying and performing calculations with data are referred to as models. Is the word model now so holy that every worker must have something which he can call *his* model?

Taaffe (1970, pp. 37—8), writing about maps as models, states:

> Maps are extremely efficient devices for the storage of spatially associated data. Although more information . . . can be stored in computers, spatial associations between the bits of information cannot readily be viewed. It has been estimated that the average US topographic quadrangle . . . contains over 100 million separate bits of information.

Not only the map can serve as a model device for storing data. Clearly the aerial photograph can serve as well, or better in

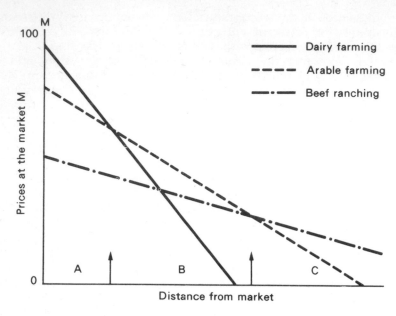

Assuming: Fixed prices at M, all inputs on farms also originate from M.
Then: The costs of inputs and transport to market increase with
distance from M, but at different rates, so that, for example,
dairy most profitable in zone A. Arable in B, etc. giving the
kind of spatial pattern seen more fully below

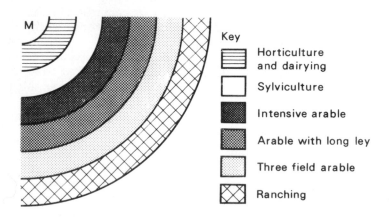

Fig. 3.1 Von Thünen's model of rural land use

certain circumstances. Moreover, the regional device can serve in a similar way for the storage of data about a definite part of the earth's surface (see Minshull, 1967, ch. 1).

Berry (Berry and Marble, 1968, ch. 2) has elaborated on this function of the regional method of description. At his most derogatory he refers to regional geography as little more than a gazetteer in which factual information can be stored and later found as required. In a more encouraging mood he does, however, expound the virtues of the regional method as a unique geographical type of classification, in which phenomena are classified together because they exist together on the earth's surface, and other attributes by which they might be classified are ignored. Thus a botanist would classify olives, vines and wheat separately, but would combine maize and rice with wheat because they are grains. The regional geographer, implicitly, classifies wheat, vines and olives together because he finds them thus in a particular region, but classifies wheat, maize and rice separately because each is dominant in different parts of the world. The geographer, and especially the regional geographer, has unique methods such as the map, the air photograph and the regional method with which to handle his data. Whether they need to be called models is open to question.

Berry, in fact, has done much to prevent regional geography from being completely ignored and discarded by the 'new', general, model-using, quantitative and above all 'scientific' geographers, by redefining it in the modern idiom. He shows how the regional approach can be viewed as a data classificatory matrix language. He shows how the questions and problems of regional geography can be redefined so that model-building and quantitative techniques can be employed and general systems theory be applied. But in spite of this valuable work, there is still something lacking which will tend to exclude regional work from the main stream of geography for the time being. This is the difficulty of setting up and testing hypotheses in traditional regional geography; possibly the absence of any worthwhile hypotheses to test.

This third type of model in section 4, that on which experiments can be carried out and hypotheses tested, is essentially the complete model, comprising parts I, II and III proposed at the end of chapter 2. A model which contains data on which experiments can be made is more than just a hardware sub-model, a map, or a piece of mathematics. Such a model must contain in addition at least some representation of the working of a system, and an hypothesis about how or why it works in that way. It is in this connection that the words 'simulation', 'stochastic', 'deterministic' and 'probabilistic' arise most frequently. For some geographers, particularly economic geographers, the model in which the data is stored and on which any experiments or tests are carried out is a model, or program, in a computer.

In this case the model may be able to simulate the functioning of the real system. The word 'may' is used advisedly, because it is not always the case that the processes *are* simulated, and the alternative 'black box' model will be explained later. The computer is valuable in several respects for this kind of work. First, as a working machine itself it can simulate or in some other way represent the working of a system. Second, as the geographer attempts to make his model behave like the real thing, it may be a case of trial and error, and the computer can make many trial and error runs rapidly, if expensively. Third, once the model represents the present functioning, such operations can be carried out on it as changing one of the factors, either to test an hypothesis about the relative importance of that factor, or to be able to predict changes in the system, say, with changes in time. Needless to say, all the data experimented on in this way must be quantitative. Data simply stored, whether in a matrix, map, photograph or verbal account need not necessarily be quantified or quantifiable.

Fourth, it is believed that most of the systems simulated in the computer behave in a predictable way. This way may be deterministic, in the sense that a given result always follows from a given cause. In this case a deterministic law may be formulated to describe the events. In contrast, the behaviour may be predictable only in a statistical sense, for example that there is only a certain percentage chance of a given result following a given cause; or that out of 100 things or people, only a certain number will behave in the expected way. In this case a stochastic law may be formulated to state the probability of effect following cause, or to state the numbers which will behave in the expected manner, with the degree of possible error in the prediction.

Much of the so-called quantitative geography and quantitative methods borrowed from other disciplines is concerned with the statistics of probability rather than with the quantitative data describing things on the earth's surface. But understanding of probability is vital to an understanding of many of the topics which modern geography investigates. Although geographers can no longer naively state, in a definite deterministic way, that men in South Wales mine coal or women in Macclesfield weave silk, the probabilities of a man finding work in the local mines and so on can be worked out. While each individual in a country such as Britain may seem to have a free choice of job, in aggregate the proportions of the 50 million people who find work in each occupation are predictable, and in total the population behaves in a logical way which can be described by the laws of probability. To clear up one confusing, if minor, point, models which are concerned with chance and probability are sometimes called Monte Carlo models. The parallel is clear enough; while the ball in the roulette wheel can stop anywhere at the end of one spin, in the infinitely long run it will have stopped an equal, and therefore predictable, number of times at each point of the wheel. Similarly in geography, while individual behaviour or the individual location may appear to be random, unpredictable and illogical, when we deal with many cases, or masses of people, then the behaviour can be predicted, with known degrees of error, by stochastic laws (see, for example, Gregory, 1971; Cole and King, 1968).

The most difficult item to deal with, in this attempt to impose some order on the types of models mentioned by many different authorities, is the map. With the possible exceptions of Taaffe (1970), and King (in Balchin, 1970, ch. 6) commentators on models in geography do not refer explicitly to the map as a model, but maps are quoted as examples of so many kinds of models, so frequently, that the position of maps in relation to models is confused, and some attempt at clarification is necessary.

Maps of various kinds are cited as examples of iconic, analogue and symbolic models. As iconic models they are said to be two-dimensional representations of a landscape; as analogues, Haggett (1965, p. 20) cites coloured lines as analogues for roads, and Yeates (1968, p. 6) contours as analogues for relief; more diagrammatic maps such as topological transformations are cited as symbolic models. No claim that maps are mathematical or natural models has come to the author's attention, but Taaffe (1970, p. 38) is quite certain that maps are experimental models, 'they may be used for testing hypotheses on spatial organisation'. At the beginning of a very useful section which outlines the many modern quantitative uses of maps, Taaffe shows a more general concept of the map as a model when he states, 'in the sense that they compress, abstract and simplify reality, maps serve as models that retain the spatial relationships and juxtapositions relevant for particular purposes of analysis' (p. 37). Later, he gives different slants on maps, 'models that permit the measurement and analysis of both static and dynamic spatial relationships', and 'devices for generating hypotheses'. Obviously Taaffe has a flexible idea of the meaning of the word model, and a strong conviction that maps can serve as many types of model.

Cole and King (1968, p. 468) stress the role of the map as a scale or iconic model, but several of their statements in the section headed 'the map as a model' indicate that they recognise circumstances where the map serves as an analogue or as a symbolic model. They, like Taaffe, stress the qualities of compression, simplification and abstraction which are both inevitable, unavoidable in making a map, and actively to be desired in making a model. However much we might want a map to show *everything* to be found in the field, it is impossible, and we accept the simplification and conventional devices of the map. On the other hand we do not want a model to contain or represent *every* feature of the original phenomenon. We abstract those elements believed to be significant, and simplify in order to comprehend and understand. So the characteristics of map and model which make the map an analogy for the model and enable the map so often to be used as a model come about for completely different reasons.

Berry and Taaffe agree that the map can serve as a procedural model for storing, classifying and working on data, although Taaffe elaborates on this last point more than Berry. In simple language the map acts as an extremely useful, general or multipurpose tool for the geographer. No apology is made for labelling 'model' what in plain language could be called a tool. The word 'model' is used sufficiently often to mean tool, particularly with reference to maps, that the fact needs to be made clear here. Taaffe, in particular, writes of the value of maps in the early stages of an investigation, but many others mention maps more particularly as useful in the final stages of presenting one's findings and conclusions. This is yet another instance of each worker being aware only of what he does. Of all the devices which have been named models, the map is probably the most generally useful, and the one with the largest number of different functions.

Referring to 5 and 2 in the list, maps obviously can be vital submodels at any stage of the research work. In Harvey's (1969, pp. 151—4) terminology they can be *a priori* or *a posteriori* models. Usually a map is likely to be a descriptive model, simply because it is a map of a real place and by definition is a graphical device to describe part of the earth's surface in symbolic terms. However, it is possible to conceive special circumstances where a map, or, more properly a cartographic diagram, has been set up to represent an ideal state of affairs with which reality can be compared. In this case it would probably be called a normative model, but that particular term will be questioned later.

One way in which a brief exposition of models can mislead is that in giving one example of each type, say, a plaster model as an example of a hardware model, the impression is given that the model and the example are one and the same thing; that hardware models are always plaster relief models, and that plaster models are always hardware models. A review of maps as examples will serve both to help clarify the position of maps, and to demonstrate that one example (the map) is not tied only to one type of model. Moreover, just as the map can be an analogue and an experimental model, maps, equations, hydraulic tanks and so on can number among the many possible examples of experimental models.

Referring to the lists in this chapter and in chapter 2, maps are examples in the following ways. As they can be used at any stage in the work, they are *a priori*, concurrent, and *a posteriori* models, according to how and when they are used. We have seen that they are also used to store, classify and work on symbolic data. A map, as a model, must be a static model, but it may well represent one state or stage of a dynamic system or of a development in time. The map may function to help generate an hypothesis, or to test one at a later stage, but with the rare exception will be a descriptive model.

Fitting the map to the list at the beginning of this chapter has been straightforward, but one must question the way in which some authorities have fitted maps to, or included maps in, the lists given in chapter 2. The most common example quoted for the analogue model is the map. On consideration, however, it makes more sense to regard the map as an attempt to *represent* the landscape, not to provide an *analogy* for it. If an artist were in the same position as a cartographer, to *represent* a tiger he would make a statue or draw a picture. His example of an *analogy* for a tiger would be, perhaps, a cat. Now the map is not the equivalent of the cat; it is not an analogy for a landscape. The map has much more in common with the statue or the drawing of the tiger. Strictly, a relief model would be the equivalent of the statue, but relief models are too difficult and expensive to make for the coverage of a whole country, and too inconvenient to use. But because the relief map would be ideal, one insists that the map is a compromise between the statue and the drawing, rather than the exact equivalent of the drawing.

If this is accepted, then there are grounds for classifying the map, not as an analogue model, but as either an iconic or a symbolic model. One cannot agree with King in classifying maps as a type of model in their own right, but, bearing in mind that the map is a compromise between statue and drawing, one understands her attempt to classify maps as two-dimensional iconic models. As a two-dimensional iconic model is almost a contradiction in terms, it seems most logical to classify the map, when it is being used to represent the structure of a system, as symbolic. It does not matter that the word symbolic more often means mathematical symbols, and it helps to emphasise that each type can have several manifestations.

In conclusion, the word 'map' is not synonymous with the word 'model'. There are some uses of the map, which, at the time of writing, have not been labelled model uses even by those who label everything in sight models. Clearly there are many models which are not maps. This partial confusion of the two devices arises from the fact that the characteristics of simplification, idealisation and abstraction are so uncannily similar, although they arise from different causes. The characteristics and uses of models and maps simply overlap to a surprising extent.

The only clear introduction of *a priori* and *a posteriori* models into geography is that made by Harvey. At the risk of introducing yet more jargon he draws attention to the fact that the model is a device used at a particular stage of an investigation. It is vital that attention should be drawn to this matter, but entirely unnecessary that an iconic model used at an early stage must also be called an *a priori* model, and perhaps the same iconic model, used at a later stage in another piece of work, must be called an *a posteriori* model. However, it would be irresponsible to confuse the issue further by rejecting or changing the terminology. The terminology exists, perhaps unnecessarily, and the aim here is to understand its relationship to the many other names for models.

The *a posteriori* model will be considered first, not because Harvey does so, but because it seems to be the more logical type of model. Ackoff's definition is some help here, in that it contains the seemingly obvious assumption that we must know about the real thing before we can model it. Can we make a model before we know the real thing? According to Harvey, the *a posteriori* model is used at a late stage in the work, as follows. Observations of reality reveal some features which demand explanation. A theory or hypothesis is proposed as a possible explanation. This hypothesis is verified and then, and only then according to Harvey, a model is made. Presumably this model represents both the observed phenomena and the verified hypothesis. It is then used to make deductions and to 'simplify calculations'. The *a posteriori* model is developed to represent the theory, for the purposes of testing it. 'In this case the function of the model is simply to represent something which is already known' (Harvey, 1969, p. 252).

The *a priori* model receives more attention from Harvey, who

(a) Theory: the model

i Elements: Cities, towns, villages
   which act as service centres
   or central places
   The surrounding countryside

ii Functioning: Each central place
    serves the area round itself,
    and is supplied by that area

iii Hypothesis: That each type of
     settlement of a given size is
     regularly spaced, in order to
     serve the countryside efficiently.
     One city serving whole area A
     several towns each serving
     part of A many villages each
     serving own small area

■ City   ● Town   × Village

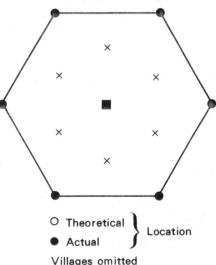

○ Theoretical ⎱
               ⎰ Location
● Actual

Villages omitted

(b) Reality: part of Lincolnshire

Town	Population
1 Lincoln	77 000
2 Gainsborough	17 000
3 Market Rasen	3 000
4 Horncastle	4 000
5 Sleaford	8 000
6 Newark	25 000
7 Retford	19 000
8 Woodhall Spa	2 000

Discrepancies
2 and 3 nearer to Lincoln
6 and 7 further north
Woodhall Spa 'extra'
Towns not of equal size
not as regularly spaced
as expected

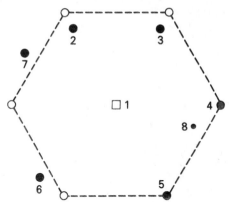

Fig. 3.2 The central place model

quotes evidence that it is much the more common of the two.
Before any particular observations have been made, before there
are any particular questions to be answered or problems to be
solved, the *a priori* model exists in an abstract form. It is difficult to imagine an abstract iconic model, or an abstract analogy
and presumably, although, again, it is not stated, the abstract

models are those of how a system could function or are models of types of explanation. For example, the gravity model equation can exist in abstract form,

$$m = \frac{PiPj}{dij^2}$$

without referring to any particular phenomena, simply suggesting the relationship of movement (m) between things at places Pi and Pj which are a certain distance (d) apart. Similarly a type of explanation that some phenomena cluster together to reduce distance between them while others are dispersed to make maximum use of space can obviously be applied to several types of phenomena studied by geographers.

So the model exists, either as an abstract logical construct or because it has been borrowed from another discipline, just as the gravity model has been borrowed from physics without referring to planets or anything else in particular. The next stage is to map in some particular data for which the model appears to be relevant. Thus the population sizes of two towns Pi and Pj may be mapped in to the equation, plus the distance or the square of the distance between them. Similarly, houses in a village or shoe shops in the town centre may be mapped in to the cluster model, isolated farms or suburban corner shops into the dispersed model. 'In other words we begin with the calculus and then seek to identify a domain of objects and events to which it can be applied' (Harvey, 1969, p. 153). Then, and only then, is the theory or hypothesis proposed to account for the observed movement, clustering, dispersal, or whatever.

Harvey makes clear both the definitions of *a priori* and *a posteriori* models and the stage during a programme of research at which they are used. The two models are not different in kind, for the *a posteriori* model of one piece of work can become the *a priori* model of another. Nor are they different in kind from any other models mentioned so far. Yet there does seem to be something missing, and it may be that in this particular section of his book Harvey considers only two-thirds of the whole, while most other writers on models consider the other third.

When, with reference to the *a posteriori* model, Harvey writes that a theory is proposed and verified before the model is made; and when, with reference to the *a priori* model, he writes that an hypothesis is proposed after data has been mapped in to the model, these very processes of proposing, testing and verifying an hypothesis are processes for which models are used or advocated by other writers. It may be only a matter of terminology.

Harvey may have in mind only what were defined in chapter 2 as submodels, of types I, II or III, while other writers refer to what was defined as the complete model comprising all three types. Another variation on the same theme is the admission in many places that the same device, or model, may serve to generate an hypothesis, to test it, and then to illustrate the verified hypothesis for the purposes of teaching.

The phrases *a priori* and *a posteriori* refer to positions in time or space relative to something else. This terminology (to be fair, not originated by Harvey) is misleading because it seems to limit the use of models to only two stages in research work. Therefore, two suggestions will be made. First, that if the terms *a priori* and *a posteriori* are to be kept, then to clarify the matter of the position of models in the stages of work, some term such as concurrent model needs to be interposed between the two. Second, because the terminology is so confusing, because the stages of work are not so clear cut, the value of this section of Harvey's book lies in the implication that models enter into the work in at least two different ways. A geographer may attempt to make a model to represent what he is studying, or adopt a model to represent what he is studying. The stage at which he does this is often immaterial. For *a posteriori* read made-to-measure model; for *a priori* read off-the-peg.

Harvey also devotes considerable space to such things as over-identified and unidentified models, and models of explanation. An attempt has been made to introduce and explain the terminology of models but these are best left until later. It will be sufficient to note here that *a priori* models in particular may not be applicable to certain phenomena, or may apply to many kinds of phenomena with equal ease, and with equal likelihood of being completely misleading. This matter of identification will therefore be considered with some of the many other dangers of models.

Models of types of explanation can be subdivided into

A. *Cause and effect models*
B. *Temporal models*
    process models
    narrative models
    models of time or stages
    models of historical processes
C. *Functional models*

The use of these models of explanation will be discussed later, and for the present it will simply be noted that such types of models exist. It has been mentioned above that in many instances a model is not expected to contain even an hypothetical explantion, whereas in other instances this is vital. Because some complete models of systems do contain explanation, there is some justification for adding Harvey's list of models of explanation to the list at the end of chapter 2. This list could then take its place as examples of submodels of explanation, theoretical or conceptual models, type III.

The situation is not quite as simple as this. Just as the complete list of submodels types I, II and III could be included under heading 4, operational purpose of the model, at the start of this chapter, so models of explanation could also be classified as tools or procedures. This particular example may be confusing, but any one familiar with classifications and attempts to classify phenomena will be well aware that any one item can fit in two or more places in the classification, according to the attribute of that item considered dominant at a particular moment. A correct classification must classify by only one attribute at a time. In the case of models as defined in geography, they have so many attributes that the same models can appear in these different classifications.

The idea behind the model of an explanation is that several logical constructs, processes of reasoning, are already in existence, and rather than start from scratch to work out not only one's own explanation, but also logical structures to test the internal validity of that explanation, one models the particular explanation on one of the general types of explanation. Thus there are several readymade procedures or tools which can speed the work and ensure more reliable results. From this we have three ideas about the model which could not be found in Ackoff's examination of the dictionary word. As a noun, model means a representation, as an adjective, model means ideal, as a verb, model means to demonstrate. But we can now see that extra meanings have been given to the word in an unexpected way. The model is a tool, the model is a procedure, the model is something on which to experiment.

Two of Ackoff's three points appear again under heading 2 at the start of the chapter, functions of the model, as descriptive and normative. Throughout most of the book so far, reference has been made mainly to descriptive models, those which in some way attempt to represent part of the world as it is. Attention is now drawn to the normative or idealistic model which represents some system perhaps in a future condition, perhaps unaffected by such things as relief and climate, or perhaps as its most perfect development. This ideal or norm is used for comparison with the real thing. Clearly, the stage has been reached where the uses of models, rather than their names and nature, need further consideration.

The student wanting to know about models rather than to use them may well be concerned to learn the names and classifications, but the perpetuation of jargon is not the aim of this chapter. The terminology has been stressed partly because it is a feature in the major works to which this book is intended as an introduction. Familiarity with this terminology can remove an obstacle to understanding the ideas behind the labels. Perhaps the best way to comprehend these ideas is to carry out a piece of research using models, but it has been noted that research in geomorphology or economic geography gives the student only a partial view of the whole, and in any case an introduction is still vital. So the terminology has been set out because it will have to be tackled, but the main purpose of the succeeding chapters will be to investigate the ideas behind the names.

4
The use of models in research

The use of models will depend, naturally, on what the geographer is trying to do. The question of the use of models has indefinite limits because they are of more use in some branches of geography than others, and because some geographers now apply the word 'model' to devices which have been well known for a long time under other names. Throughout this chapter it will be assumed that models are aids to research, and in the next chapter that they can be aids to exposition. In both chapters the major part of the attention will be not only on general, systematic geography, but also on that type of general geography which is concerned with spatial organisation. This is not to deny that models will be useful in regional geography, but there is little evidence of their use there as yet.

The initial assumption, then, in this chapter is that research is being carried out in the separate branches of general geography. As these *are* defined as branches of geography, and not as geomorphology, climatology, economics or sociology, it will also be assumed that the geographers carrying out the research are concerned to understand the arrangement on the earth's surface of the phenomena in question. They, in their turn, make the assumption which is the common element in the new geography (Minshull, 1970) that there is a logic behind this spatial arrangement and spatial interaction which can explain the patterns and functions we observe.

It is essential to understand this change in the emphasis of geography if one is to understand why models attract so much attention and why they can be so useful in the new types of work. The study of rural settlement furnishes a useful example. The phenomena associated with rural settlement have always been of interest to geographers. It is not a case of geographers having been interested in one aspect of rural settlement in the past, and a completely different aspect now, but the emphasis on different aspects has changed to a marked degree.

In the past rural settlement was described in detail, and in particular terms, with frequent reference to specific examples and places. Then, as now, there was great interest in the contrast between nucleated villages and dispersed, isolated farms. Certainly reference was made to theories that nucleation was related to arable farming and dispersal to pastoral farming, but the explanation of such patterns was usually in terms of how a particular example had come about. In a sense, the main explanation was historical; given a certain pattern of villages in a specific region, a specific explanation was offered as to how that pattern had developed in time, sometimes with a regional change from nucleated to dispersed, or vice versa (see Houston, 1970).

Nowadays less energy is spent on description, and the interest is in types of rural settlement more than in specific regional examples. There is still as much interest in the contrast between nucleated and dispersed patterns, but little in how they have developed over the centuries. The modern geographer's attention is turned to how such settlements function in space now, and how far they are good or bad responses by man to the problems of using space and overcoming distance.

Thus the nucleated village, above all, is seen as the type of settlement which best serves man's social needs. If, moreover, each farmer's land is divided into scattered plots, or if all the farmers pool their labour, then a village located at the centre of the scatter of plots, or at the centre of the communal land, is the best *spatial arrangement*. The dispersed settlement with isolated farms makes least provision for man's social needs. But if one farmer has all his land in one block, if he operates his farm as an independent commercial unit, then the location of the farm house and farm buildings on his block of land makes most economic sense.

Historical explanations of the pattern of rural settlement have proved unsatisfactory to the geographer. For example, one pattern has not proved to be linked with one group of people or one type of farming. Saxons have produced both dispersed and nucleated settlement, while Danes and Celts have produced patterns sometimes claimed to be typically Saxon. Some societies have adopted villages as centres to grow cereals, as in Europe, others have established isolated farms to grow cereals, as in the United States. What interests geographers now is the attempt to work out a satisfactory spatial system on the earth's surface, no matter by what race, or for what particular crop. The elements of the system are the fields, the farm house, other farmers and their houses, and the market. Things such as cash, food, goods, fertiliser, seed, animals, labour, machinery and power come into the system, often from a central place which is also the market. The farmers then arrange their fields, houses, paths and roads to make use of the land, labour and other inputs. This is the functioning of the system, combined with the flow of the crops and animals produced on the farms to the market, the outputs of the system. There may be several explanations for the working of the system, and for the arrangement on the earth's surface of fields, farms, roads and market towns. The explanation may lie in the farmers' attempt to maximise their profits, or to maintain a desired social system, or to husband the soil for future generations. Having identified the elements and the basic functioning of the system, the geographer should look for a satisfactory explanation of *why it still functions like this*, not how the outward form is a legacy from the past.

A short digression into the past situation will throw light on the present problem. Some use of models *was* relevant to this earlier work. In order to describe the pattern of rural settlement the descriptive geographer did in fact have to simplify, abstract and compress his material. In the case of rural settlement, the subject was often simplified until it was a matter only of points and clusters of points on the earth's surface. As settlement was the topic of interest, it was abstracted from everything else in the landscape considered irrelevant. It was partially abstracted from such things as fields, crops and communications, and completely abstracted from such things as manufacturing which might also exist in the same landscape. Thirdly, whenever a number of villages larger than the geographer could see from one vantage point was included in the study, then compression into an artificial bird's-eye view took place.

Description, simplification, abstraction and compression were achieved by a verbal account and a map. The map simplified the settlement into symbols by showing only the few things which the geographer wanted to show; it abstracted certain features from the complex of phenomena on the earth's surface, and by representing many square miles on a piece of paper a few inches square, compressed the arrangement to a manageable size.

The map and the verbal account suggested so far would cover mainly the structure of the system. The functioning of the system — the way the people arranged their crops round their residences, the types and frequencies of movement of men and goods both within the system, and between the system and market — was more likely to be set out in verbal terms than in any other way. This state of affairs enabled modern critics to begin to write of an irreparable division between the old qualitative geography and the new quantitative. However, there were attempts to quantify some aspects of the system, for example Demangeon's (1933) formula for the coefficient of nucleation:

$$K = \frac{E \times N}{T}$$

where K is the coefficient, E the population of the commune excluding the largest settlement, N the number of settlements minus one, and T the total population of the commune. Forty years ago the geographer, at this stage, would think in terms of having a description, a map and a formula, in modern jargon he had a descriptive model, a symbolic or iconic model, and a mathematical model.

Further, when some explanation was offered as to how the particular pattern of rural settlement came about, the geographer would possibly use what Harvey and others now call models of explanation. If the explanation followed deterministic lines, as fashionable in their day as models are now, arguing that the limestone rock, barren plain with little surface water and only patches of soil 'gave rise' to nucleated settlement, in modern terminology this would be using a cause and effect model. If, on the other hand, the story of the settlement was given, from the day the first people arrived, this would be using a narrative model. Again, it may have been assumed that all settlement passes through the stages of farms, hamlets and villages, and a model of historical process might have been used. In any case, it is unlikely that the word 'model' would have been used either as often or as selfconsciously as it is today.

The pattern of study of rural settlement as outlined here would receive harsh criticism from certain geographers today on four counts. First, because the interest was in how things came about rather than how they are arranged and function in space. Second, because it usually did not try to develop models. Third, because it rarely quantified its data; sometimes the very interest in an historical rather than a functional explanation precluded quantification, but at other times attempts were made, laying the foundations for modern work. Fourth, because it did not test hypotheses in the way demanded by the modern scientific approach.

One cannot argue about the aims of other geographers and say what they should have done or ought to do. If they aimed to give an historical explanation, that is that. But in giving this kind of explanation as distinct from any other possible kind, they laid themselves open to difficulties which face all historians. Here we can learn a valuable lesson. Above all, they were putting forward hypotheses which it was often impossible to test and verify.

Too often an hypothetical explanation in history, and in this type of geography, is never even seen for what it is — an hypothesis which must be tested and either verified or rejected. Two factors help to bring about this state of affairs, the fact that the hypotheses can sound so plausible and acceptable, and the fact that the historical evidence needed to test them is so often completely missing. Houston (1970, pp. 85—90) gives a review of the theories on the origin of settlement types, revealing a situation where there are conflicting theories, yet each plausible in its own way and each impossible to verify or reject completely. For example, Meitzen's theory that dispersed settlement is Celtic, nucleated settlement Germanic, and round and street villages Slavonic is contradicted by Flach. The 'Roman' school of thought which insists that settlement patterns in England are continuous since Roman times is opposed by the 'Teutonic' school which insists on a complete break at the time of the Germanic invasions. Similarly, Gradmann argued, most plausibly, that Neolithic settlers, being unable to clear the forests, had to settle on the poorer soils which supported less dense vegetation; but Nietzsch argued, equally plausibly, that Neolithic settlers depended on the forest, as an indicator of soils, for raw materials, for protection, and to feed their animals.

As further evidence of Neolithic sites and of the type of vegetation in the areas where they settled, at the time they settled, is accumulated, this argument may be settled. But many others will remain as conflicting hypotheses, each of which attracts adherents who can 'see' that theirs is the correct explanation and that their opponents are obviously wrong. Whether some writers are aware that they are only setting up hypotheses, or whether they genuinely believe they have explained something must remain a mystery. What is certain, is that many of their students and readers at once accept the hypothesis as verified fact.

In all branches of geography, in addition to the study of rural settlement, and distinct from history, the same mistakes have been made. Original workers often have not made clear the hypothetical nature of their explanations, while others have been much too ready to accept the cycle of erosion, a theory of the location of industry, or an hypothesis that man is determined by his physical environment as self-evidently true. So the fourth, and major criticism of much traditional geography is twofold: that hypotheses often have not been set up in such a way that they can be tested; and even in those cases where they have been set up properly, they have not been tested rigorously.

In order to examine the uses of models, which also at times involve the use of quantitative techniques, it is necessary, as argued above, to accept that much explanation is of an hypothetical nature. Moreover, it is necessary to accept that the use of models makes much more sense in one type of geography than in another. It is not necessary to insist that other types of geography are worthless and redundant. Once these conditions have been accepted, then the use of models in geographical research can be examined under six headings: as aids to represent systems, as a stage in theory construction, as a device to separate laws from generalisations, as simulation devices, as experimental media, and as tools.

1. Models as devices to represent systems

This use of the model has been the burden of most of the argument thus far, and so needs little further attention. As soon as a geographer turns his attention to some phenomena on the earth's surface which he believes to have a purposive distribution or to have functional interaction, inevitably he begins to model them. He never simply observes and handles the real things, farms, factories or towns, he represents them in words, maps, diagrams and mathematics, and studies and works on these representations.

It is possible to represent the elements and interaction of a system without explicitly trying to explain why, say, the factories are distributed that way, and why certain goods are moved between them and other places. Some geographers certainly seem to stop when they have gone this far. But a more complete or elaborate model can also contain one *or more* hypothetical functional explanations for the distribution and interaction which have been observed. The phrase functional explanation is used in the way described by Harvey (1969, chs 20—2) to distinguish this type of explanation from a temporal one. The essential features of such a complete model are that it should represent elements and functional relationships which can be identified unambiguously by other workers, and set out the suggested explanations in such a way that they can be tested and verified or rejected, by both the originator and other workers.

The first part of this use of a model is neither impossibly difficult, nor new. Many geographers have modelled a man—land system in which both the elements and the key inter-relationships have been identified. Such elements are the rock, relief, climate, soil and wild vegetation of a particular area, the farms, crops, houses and people who inhabit it. The functional relationships of man taking minerals and water from the rocks, clearing the vegetation and using the soil, have been described.

But here the error has crept in. The effects of high relief forcing man to terrace the land or go in for pastoral farming, of the climate making him grow certain crops, use a certain type of architecture and clothing and so on, have seemed to be described when in fact they were only being hypothesised. The argument seemed to run, these are the features, these are the actions which take place, and they take place because physical features P determine the human actions H.

Then, as now, the argument should be stated: these are the features, these are the actions which take place, and it is suggested that these are the crucial features and the actions are the connecting actions because of a possible cause-and-effect connection between P and H. Then it must be stated how the hypothesis can be tested, under what conditions, and what results would indicate verification or rejection. It is mainly because the problem has never been stated in such terms that the argument about determinism, probabilism and possibilism dragged on for so long.

Perhaps this extreme case should be modified. There will be instances where the testing of the model suggests a change in the model, rather than complete rejection. Providing the modified model still fits in with what the geographer is trying to do, and he has not unwittingly changed his ground, this is acceptable. It is often argued that the most fruitful way to develop reliable theory is so to construct the model that several hypotheses can be tested at once. This may well be the best method, but it is one which is rarely seen in action. Too often the geographer is concerned to test his favourite hypothesis, and in the extreme case, to prove himself right.

A similar difficulty is encountered when an hypothesis which was sufficiently verified to be acceptable for a time has to be rejected later because a better hypothesis is put forward, or new facts which have come to light demand a different explanation. One then witnesses the embarrassing process of the old hypothesis being vilified and denigrated by the champions of the new. The new champions, in their turn, fail to accept that their explanation *is* only an hypothesis, and it, too, will be acceptable only until such time as new facts, different cases or better explanations necessitate a change. W. M. Davis's hypothesis that the elements of the landscape we observe have been produced by the processes of river erosion developing through a complete cycle from youth to old age served very well in its time. In fact it served so well that it was for a long time accepted perhaps too uncritically by too many people. The odds are that whatever replaces the concept of the cycle of erosion will not be the final and complete explanation.

2. The model as a stage in theory construction

A basic assumption behind much of the use of models is that geographers nowadays are working towards general theories rather than concerning themselves with specific places on the earth's surface. Thus the use of models is part of the whole change in the paradigm, away from regional, subjective work and towards general geography and the scientific method. The use of a model as a stage in the construction of general theory therefore has two aspects.

First, as the general geographer is interested in the geography of one phenomenon, whether landforms, farms or towns, as distinct from the geography of all the phenomena in one place or region, he is concerned to ignore unique regional accidents in the distribution of his phenomenon. Thus the general urban geographer not only ignores other phenomena in a region, he must also study cases of towns in many places, in order to determine their general characteristics. Models which simplify, abstract, compress and put the emphasis on what are believed to be the important elements of a system greatly facilitate this sorting out of the general from the particular.

A crude example can be given of towns as systems made up of different functional zones. One town may have a zone of pump rooms, hospitals, clinics and convalescent homes which functions as a spa. Several other towns will have zones which function as ports, yet others which function as holiday resorts. But as each town investigated is reduced to a model of functional zones, observed in the field and recorded in a list or on a coloured land use map, and as more and more towns are observed, it will be seen that every town has a central business zone, one or more residential zones, transport and manufacturing functions. Spas, ports and resorts can then be seen as special cases, particular regional manifestations of the basic town, each in addition having its business, residential, transport and manufacturing functions. If general theory is to be developed, it must be based on those elements which are common to every case of the phenomenon in question.

Second, Harvey (1969, p. 142) uses the word theory to mean something different from the word hypothesis, in effect to mean an hypothesis which has been verified and accepted. Once the model has been used to help sort the general from the particular, it serves to test the hypothesis, so that if verified it may become general theory.

Here we have touched on one of the most important and one of the most difficult uses of the model. Much of Harvey's attention in *Explanation in Geography* is devoted to the development of general theory in geography and to the role which models can play in this work. The necessary detailed considerations are beyond the scope of an introductory work such as the present volume, and the interested reader is urged to study Harvey's work very carefully.

One word of warning can be given in this connection. Possibly because of imprecise terminology, at times one is given the impression that the model *becomes* the theory. This is not so. It will become clearer when simulation, normative and ideal models are considered below that models only *represent* theories, just as they represent elements and functions of systems. Once theory has been developed and established, virtually the same model device may be used to represent the theory, or give examples, for the purposes of exposition and teaching. Harvey keeps hammering away at this point, and it takes time to appreciate its importance. Harvey would be much easier to understand if he used examples to demonstrate his points. Just as we would not mistake an analogy for the real thing — mistake a cat for a tiger — we must not mistake a model for the real thing or for a theory about it.

3. The model as a device to separate laws from generalisations

Several teachers on first acquaintance with the 'new' models have rejected them out of hand as being new jargon for the old generalisations which teachers and geographers have always made. It is this kind of reaction which gives one grounds for insisting that if models are really something new, and something really worthwhile, they must be models which comprise all three of types I, II and III and must include the type of explanation which can become law.

Geographers have always made general statements about phenomena on the earth's surface, and if models are no more than a revamping of this process they are open to exactly the same kind of criticisms as the old general geography. In particular, the criticisms are that the generalisations were qualitative rather than quantitative, were unproved, and were formulated in such a way that they could apply only to cases which had already been studied and were known. So the implication is that the new models, if they really are something better than the old generalisations, must be quantitative, verifiable, and, above all, universally applicable. In this way the general geographer has some hope of establishing general scientific laws, rather than of making generalisations about specific examples.

When a specific example has been examined in detail and is known, for example a town, then any description or explanation of the interrelationships of the parts of the town can refer only to that town. When many towns have been investigated and the relevant data collected, then general statements, generalisations, can be made which refer to all the towns which have been studied, *but to no others*. This is, and has been, common practice at all levels of teaching. Particularly in regional geography generalisations are made about the vast number of towns which are known and have been investigated by countless field workers, and examples are quoted to illustrate specific points. The teacher may be able to generalise, with some confidence, that every town in North America or Europe has a central business district, perhaps a generalisation which some readers will consider hardly worth making. The teacher, however, cannot say that every town, in any place and at any time must, *of necessity*, because of the very nature of the thing called town, have a central business district. If he could state this, he would be doing something more than extend the general statement to cover cases not yet examined by geographers, he would be making a different kind of statement about the nature of towns, he would be stating a law.

We can generalise about all the rivers we have seen, and say that all those *do* run in valleys. But to state a law which will apply to all rivers we have never seen, all rivers in the past and in the future, we must hypothesise some way in which rivers *must* run in valleys of necessity. Very often, children do not realise this until the fifth or sixth form in school. Before this time, they appear to think that the earth's surface just happens to be well provided with valleys, and that rivers run in them as a matter of convenience. Whether their formal instruction on landforms derives from Davis, Penck or some other authority, the assumption is that rivers of necessity flow in valleys because the rivers are one of the main agents in the process of erosion which produces those valleys.

The regional geographer cannot generalise or state laws about regions he has not examined. The general geographer of landforms is able to state laws which, because of the way they have been formulated, apply to all the cases he has not examined. More, the general geographer does not want to examine every specific case, as does the regional geographer, and therefore aims to formulate laws in the first place. This aim, of course, is scientific, and the method employed is one of the many variations of the scientific method. Scientists and general geographers were using the method before the word 'model' came into common usage, although it is now used much more widely in the sciences than in geography. Whatever the name, the attempt to identify the elements of a system, in this very simple example the river and the valley, to identify the process, erosion, and to hypothesise a *necessary* connection so that one is the function of the other, can lead to valuable, lawful statements which are of a much higher order, and greater intellectual significance, than mere generalisation.

4. Models as simulation devices

The word simulation is being used in geography with greater and greater frequency, and more diverse meanings as time goes on. Simulation games are being introduced, based on the operational games used to train military personnel and business executives. Presumably the word simulation in this context means to simulate the decisions which have to be made in commanding a battle, deciding a firm's policy, or locating a factory. It is clear, in contrast, that the word simulation in other contexts means something much less similar to the real thing, and it is in this sense that it will be used here.

Ambrose (1969, pp. 217—25) gives one of the clearest simplified introductions to the use of simulation models. He makes clear that, above all, simulation models are intended to be used for prediction of trends and events. This predictive use of models was mentioned several times in the first chapter of the present work, it being noted that workers in hydrology, demography and so on see this as one of the functions of their models. In this section the emphasis is on prediction as the main function of the model, and therefore it is useful to have a name such as simulation models, to distinguish them from others. The function of most of the models considered so far is to represent and explain; the function of the models under consideration now is to simulate and predict.

The concept of simulation must be understood very precisely in this use of a model. One might argue, with complete validity, that an hydraulic tank with a miniature river, running down a miniature valley, and depositing a cute little delta on the laboratory floor is a simulation model of river deposition. Unfortunately few model users would apply the word simulation in this way. It is much more likely that to simulate river deposition, they would devise a computer program which worked in such a way that when the computer was fed with the information about how much rain fell over the river basin in twenty-four hours, it could produce a quantitative answer about how much debris would be deposited and by how much the delta would have grown.

Obviously the computer program, or the memory, must contain accurate, quantitative information gathered by field measurement, on the rates of erosion for a given rainfall, and the rates of deposition for a given amount of erosion and volume of water in the river. When the computer, in a few seconds, simulates a week's erosion and deposition, only two of the three stages are identical. The input for the computer is the same as the input for the real river system: the amount of rainfall, the one in figures, the other in raindrops. The output from the computer is the same (one hopes) as the output from the real system — a set of figures defining the new size of the delta, which could be measured in real life. But what goes on in the computer is nothing like what goes on in the real river valley or even in the hydraulic tank. The internal processes of the system are 'simulated' but not directly copied and reproduced.

To work out a program which, for given inputs, will produce the correct outputs, may seem tedious and pointless when the real river can be observed, or a hardware model can be made. However, the real river may be inaccessible and difficult to measure, and usually the processes are tediously slow. Measurements *must* be made, but these are samples at certain places and intervals of time. Similarly the hardware model poses difficulties of reproducing both the materials and the processes in miniature. For example, sand in a scale model represents boulders in real life. Therefore constructing the computer program, the thing which is referred to as 'the model', however difficult and time consuming it may be, is often regarded as worthwhile even in this kind of investigation.

In human geography the case for a simulation model may be easier to accept. Human interactions are not directly observable on the ground, as is a delta, so they need to be represented or simulated. It is not as easy to carry out measurements of movements of people as of river flow, and one cannot experiment on people in real life, as one might divert one of the distributaries on a delta. Equally obviously, it is not possible to construct a hardware toytown, with miniature people, as one can construct a miniature river basin. So, for example, if one wants to simulate the growth of a town, for given numbers of immigrants, and to predict what will happen as more people come in, the computer simulation model at present is the best device available to human geographers.

Morrill (in Ambrose, 1969, pp. 258—79) describes in detail the use of such a computer model to simulate the growth of the Negro ghetto in Seattle. Stage one was to write, rewrite and write again the program so that it would simulate what had already happened. Eventually the computer, fed with the numbers of Negroes who moved into Seattle each year in the past, reproduced to Morrill's satisfaction the growth of the

71

ghetto to the correct size and in the correct direction. Within acceptable limits of error, Morrill then believed he could use this model or computer program to predict what will happen in the future. Prediction could take one of several forms. He could go on feeding numbers into the computer as if the present trend of Negro immigration will continue into the distant future, or he could feed in double, quadruple numbers to see if this would result in some unexpected, unforeseen development, like the growth of a second ghetto, or a completely Negro city centre.

The simulation model which is used for the purposes of prediction is sometimes called the 'black box'. This piece of jargon will be examined in chapter 6. It is stressed that prediction seems to be the main use of such models, and, moreover, their most appropriate use. It may be argued that prediction is not the geographer's job, although many people who call themselves geographers would disagree. By definition a geographer is concerned with actual phenomena, arrangements and processes on the earth's surface, not with gazing into a crystal ball or a computer printout to guess at the future. In contrast, the hydrologist, meteorologist, economist and demographer may well need to predict, and eventually help to control, the processes which they study. As these and other disciplines are so closely connected with geography as to be mistaken by many for parts of geography itself, it is also contended that the geographer should be aware of the aims of these disciplines and the modern tools they use.

The computer simulation model, however well it works, *by itself*, does not explain why a river deposits a delta or why the Negroes live in ghettoes, and in a ghetto spreading northeast in Seattle in particular. Such a model can work, and serve the purposes of prediction, *without* an explanation of the internal processes, providing that for a given input it produces a given output. Hence the unexplained black box idea in between. As Chorley (in Berry and Marble, 1968, p. 42) makes clear, the computer models have only internal logic. If the propositions are correct, then the mathematical processes will give the correct results. But these processes and results do not explain the real world, they simply help us to predict what will happen, not why it works that way. In a similar fashion, for centuries farmers have been able to predict future weather from present conditions, without in any way understanding the processes which connect the two different states of the weather at different points in time.

Harvey's (1969, p. 152) distinction between the model and the theory, and between the model and reality runs along similar lines. In this context he states that while the theory must contain an explanation of the causes of the processes we observe, the model can simulate, test and demonstrate only the results and effects. If a model represents or simulates reality to our complete satisfaction, that is all it does. The fact of excellent representation or simulation does not, *in itself*, prove that the hypothetical explanation is correct.

Ambrose (1969, p. 224) is slightly more optimistic: 'The ability to make an accurate prediction about a set of events is one indication that the events are understood.' However, he follows this with a long, cautionary paragraph, reinforcing what Chorley and Harvey say, and concluding, 'observing a correlation is not the same thing as understanding the causation'. Morrill (in Ambrose, 1969, pp. 258—79), Yeates (1968, p. 53) and many others who use simulation models, claim that they are forced to understand and explain the processes in two ways. First, in order to simulate what has happened, what does happen, accurately, as a basis for prediction, they must make very careful, detailed field observations, both to collect data for the computer and to hypothesise some mechanism. Second, during the tedious process of trial and error to make the computer program fit the observations, they claim they are forced into a greater and greater understanding of reality. Perhaps, but this seems more like spin-off than an integral part of the predictive work.

Finally, a comment on the value and reliability of prediction from the *Guardian* leader, 19 May 1971.

> The Royal Commission on Population, which was set up in 1944, had to examine fears of a falling population. Its report published in 1949 forecast some further growth in the population, but thought it would not be large. Little more than twenty years later expectations are quite different. The Commons Select Committee on Science and Technology has been told to expect the population to increase by as much in the next thirty years as it did in the past seventy. That may turn out to be wrong too.

5. Models as experimental media

By now it should be clear that the uses, separated for examination here, in fact overlap each other to some extent. In a sense, experiments are carried out on simulation models, and on representative models in the stages of theory construction. Because of this, much has been implied in earlier sections about the use of models for experimental work, but two aspects merit more direct attention.

In 1951 Martin could state, without expecting to be challenged, that geography was not an experimental science.

Whether or not it ought to be is another matter, but there is no doubt that the increasing use of models in geography, and the more determined and explicit attempts to employ scientific method, have changed the situation considerably in the last twenty years. Martin, referring to physical environmental determinism, meant specifically that geographers could not go out and experiment on groups of people in their natural environments. Now that geographers can model the societies and the environments, albeit to a limited extent, they can carry out experiments to test hypothetical explanations, and to examine the nature of the systems under observation.

The former group has received some consideration already. Once an hypothetical explanation of a system has been offered, deductions must be made and experiments carried out, preferably in the field or on direct evidence, to verify them. For example, say, it is hypothesised that central places or market towns are evenly spaced in order to serve and be served by the population in the surrounding countryside. Deductions from this are (*a*) that the town serves all the people in the area, and only people in the area, for weekly shopping; (*b*) that the farmers in the area all supply the town with food and only that town with food. Questionnaires completed in the field would soon seem to validate deduction (*a*), but evidence from farmers, combined with trade statistics, would cause a rejection of deduction (*b*). The model has been tested experimentally, and the hypothesis needs, at least, modification.

This testing in the field, or against evidence in maps and statistics is probably familiar to readers who are not aware that they have used models, but have followed a sixth-form or college course in geography. The second type of experiment, to examine the nature of systems, is less familiar, not only because it is new to geographers, but because it takes longer, demands more expensive equipment and procedures, and involves more precise reasoning. This type of experiment is carried out on a model of a complicated system, where many factors are believed to influence the operation of the system, in an attempt to determine the relative importance of the factors.

Such a model and an experimental procedure would have been invaluable years ago to investigate the workings of environmental determinism. If, and those who appreciate both the complexity of the determinist concept and the extreme difficulty of making good models will realise this is a very big if, the main physical, economic and social factors operating to produce the economy of, say, the Eskimos or Pygmies could have been represented in a model, and each factor isolated in turn, as modern models attempt to isolate them, then the importance of physical factors in determining man's way of life could have been evaluated with at least some attempt at precision. In the past, with less sophisticated techniques at their disposal, the determinists may not have been able to test their hypothesis, or investigate the systems of man—land relationships which they tried to understand. Now that some experimental techniques are coming into use, it is incumbent on the modern geographer at least to consider carefully whether such experiments can help him to be less subjective.

Obviously, if experiments are going to be carried out on the internal functions of the model, it must be much more than a black box, it must be the most accurate representation of the real system which the geographer can devise. Often, the geographer will want to measure, calculate or operate on some aspect of the model which is inaccessible to him in real life. This may involve things as different as the valley floor underneath a glacier and the emigration of young people from a Swedish town.

In the case of population pressure and emigration there will be many factors operating. The geographer may want to trace all of these, to assess their relative importance, or to assess the effects of changing the relative importance of the factors. Needless to say, both for speed and accuracy, the ideal is to be able to quantify all the data and carry out the experimental operations in a computer. The advocates for the introduction of quantitative procedures into the teaching of geography are incredibly sanguine about the ease of doing this (see for example Chorley and Haggett, 1970). The research workers who give details of their work are either rueful about the difficulties, or disarmingly frank that they couldn't quantify, or couldn't get the data they needed, so used some other factors instead (see examples in Ambrose, 1969; Berry and Marble, 1968).

However, these are early days, and perhaps one should look to the ideal, rather than the less than perfect first attempts. So it will be assumed that factors such as rates of population growth, standard of living, demand for food, available labour force can be quantified. Then, as Wrigley claims (*MG*, p. 211):

> The equations expressing the relationship between rates of population growth, trends in income per head, the structure of demand, and the fraction of the labour force employed on the land . . . make it possible to explore the effect of changes in one or more variables upon the other variables in the system in a coherent and logical fashion.

In addition, one must not underrate the difficulty of finding the right equations for the computer.

One of the most commonplace statements made about models, repeated time and again, is that they are simplifications of reality. McCarty and Lindberg, (1966, p. 55) writing of experimental models, give a hint that this is not always the case. In that the mathematical, experimental model can deal with varying amounts of the given forces, have the component factors rearranged, inserted or taken out, combined or separated, it is something at once more complicated and more all-embracing than any *one* real-life case of the phenomenon being studied. Where many factors can operate in a type of system, any one example of that type of system may be affected by only some of the factors. But the model must be able to represent and deal with them all. Thus the mathematical model can have a mass of components which are put in, left out, varied in magnitude, to fit specific cases, almost like the handyman's power drill which can have circular saws and screwdrivers attached and removed at will. For McCarty and Lindberg, a model is 'an hypothesis . . . stated in such a way that the effect of changes in any of its independent variables can be assessed' (p. 54), a limited definition, but one which emphasises the experimental function.

Nystuen (in Berry and Marble, 1968, p. 36) goes to the other extreme from McCarty and Lindberg. Admitting that his aim is to make geography as abstract as possible, he writes:

> The elements in an abstract [model] system possess only the properties explicitly assigned to them. In the real world, behaviour of a variable is often due to causes not included in the explanation because they were not thought of.

He seems not to be concerned to include factors not thought of, when he continues, 'one can restrict the properties of the object under study to a bare minimum and allow only simple associations to exist. By doing this the problem may become simple enough to understand'. Nystuen's overt purpose may be to simplify in order to understand, but in the process he, and other users of experimental models, is doing what the natural scientist does in a laboratory experiment — so controlling the elements and factors in the system that the function of each may be ascertained.

Other writers echo Nystuen in stressing the importance of simplification in the early experiments at least. Even McCarty and Lindberg admit that oversimplification is often necessary just because, as Nystuen says, we do not know, have not even thought of, all the possible factors operating in a system. For McCarty and Lindberg (p. 54), the simple models are early steps in a long journey towards discovering all the relevant factors and their magnitude. Until then, such simplifications as the isotropic surface, crude as they are, are necessary. This necessity, in the early stages, of course, leaves some of the models, and the use of models, wide open to criticism, particularly from those who will not see that they are at an early stage of development. For example, the central place model may not provide an entirely satisfactory explanation at the stage when it *has* simplified and abstracted the known, obvious factors. Some workers, not necessarily on this model, may give critics grounds for believing that they are not concerned with what J. S. Mill called 'the insufficiency of the obvious causes to account for the whole of the effect'.

6. Models as tools and procedures

The use of models to store and classify data has been outlined in chapter 3. Enough has been said about the use of models as tools for the moment. The model as a tool for the analysis of specific examples properly belongs to the next chapter. Many procedures, however, have only been mentioned, and deserve more attention here. Harvey (1969, chs 17—22) devotes most attention to model procedures in geography, although economic geographers often use the word model as though it were accepted that it means a procedure, such as a set of instructions for a computer to follow (Lowry, in Berry and Marble, 1968, pp. 53—66).

Many procedures have operational rules, and Harvey argues, sensibly enough, that the geographer will both save time and avoid error if he uses one of these procedures rather than trying to work out his own. Thus there are standardised procedures for:

Selecting data	Sampling procedures
Collecting data	Fieldwork procedures
Defining data	Ostensive, lexical and operational
Measuring data	Nominal, ordinal, interval and ratio
Manipulating data	Quantitative techniques
Classifying data	Logic and set theory
Representing data	Mapping procedures
Explaining data	Temporal, functional, cause and effect

Those who have struggled to draw good maps, have been infuriated by a bad map, or had a fieldwork sample turn out to be biased may readily understand the need for well thought out and tested procedures when they go on to less familiar operations. These procedures can greatly ease the work of one geographer, but they will have infinitely more value if and when they are used by the majority of general geographers, for then the work of each will be that much more readily available to, and understandable by, the rest. Imagine that three different people, using different procedures, have drawn maps of Cheshire, Derbyshire and Lincolnshire. Each man would have been saved time working out his method, and the maps would be much more use to many other people, if the three cartographers had used the same scale, the same symbols and the same method of showing relief.

One of the most serious questions, which exercises many philosophers, is 'What constitutes an explanation?' While Harvey (1969, chs 20—2), like anyone else, can not give a final answer to this question, he does draw our attention to the basic types of explanation at which we aim, and procedures to follow. This is a most valuable section of his book, and the models of explanation, like many others, can serve to make geographers stop to think exactly what they are trying to do, what they want to explain, and how best to set about it.

If the geographer aims to explain certain phenomena by reference to cause and effect in space, or in time, then there are rules of logic for him to follow, or against which to test his argument. Similarly, although at present much more rudimentary, there are models of functional explanations. When developed, these may prove to be the most use to general geographers who are concerned to explain spatial arrangements by reference to the way the elements of a system interact in space.

Harvey gives most detail of temporal models of explanation, mentioning in this context that many people believe that if they have described how a thing came about they have, in fact, explained it. Unfortunately this is not the case, if only for the fact that they assume that what *did* happen was the only thing which *could* have happened. Any explanation, whether cause and effect, functional or temporal, is an explanation only if it can demonstrate that the state of affairs is a matter of necessity, and not just of chance or coincidence. Thus a temporal explanation, an historical account leading up to the present state of affairs, is an explanation only if it covers all the possible chains of events in time, and demonstrates that the chain of events which did take place necessarily had to take place in this way, and in no other. However, while such an explanation would be perfectly satisfactory from an historian, even a temporal explanation which does demonstrate logical necessity is not a satisfactory type of explanation coming from a geographer. If geography is a discipline different from both history and science, it must not only examine the relationships of events in space, but also attempt to give a spatial type of explanation, to demonstrate that these elements are spatially interrelated in this way on the earth's surface of necessity, whatever their particular historical development may have been.

5
The use of models in exposition

The uses of models to be considered in this chapter are somewhat artificially separated from those in chapter 4. To a great extent these models are relevant to teaching and exposition rather than to research, but this is not to imply that the two processes take place at different times and in different places. Moreover, while the processes of research and teaching certainly do take place, it may be begging the question to assume that devices used as models in research are still models when they are used to aid demonstration. Some authorities restrict the use of the word 'model', others urge the use of models in teaching. So it will be assumed that some of the devices used in research are used in teaching and are still called models, although it is accepted that some will dispute this.

Many of those engaged in studying systems on the earth's surface from a geographical point of view aim to construct the most accurate model possible. This can be so difficult that the working life of many geographers can be devoted to one system and still not achieve a completely satisfactory model. In an ideal world, one might fairly expect that a model will not be used for the purposes of teaching, or for the purposes of examining specific cases, until it is very near to perfection. In this world, particularly with the present climate of opinion in some branches of geography, models are put to such uses at very early stages in their development. There is no point in bemoaning this state of affairs, it is a fact, and one must simply be constantly aware of it. In this discussion it will also be assumed that a very few near-perfect models, and many rough and ready models, are used for teaching and exposition. In the few cases where a model is used without any intention of it ever being perfect, that point will be made clear.

In *Frontiers of Geographical Teaching*, Haggett writes of the model as, 'an idealised representation of reality to demonstrate certain of its properties' (p. 106). He probably does not mean 'idealised' in the sense of an ideal model, to be explained later in this chapter, but he is clear that models have a function in

demonstration and teaching. It will be argued that this representation for the purposes of demonstration can contain less or
more elements than the real thing, as a simplification or as a
compendium. The functions, used as much in teaching as in
research, of using the model as a norm and as a system of
reference, will then be discussed in some detail.

7. The use of the model as a simplification

The need to make simple models of complex systems, either
because the systems are difficult to comprehend, or because the
full complexity has not yet been realised, has been stated in
connection with research. However well a complex system is
understood by certain authorities, on the other hand, there is a
need to present a simple model of it to those students who are
just beginning their first attempts to master the complexities.

It is not claimed that any simplification a teacher makes for
the purposes of introducing a subject should be given the name
of model. If that were justifiable, then our primary schools
would be the most ardent and convinced users of models in the
country. We are all too well aware that our earliest expositions
of some subjects to young children amount to distortions, as
well as simplifications. The kind of simple model in mind to be
used with students entering a recognised discipline is either an
early version of a model which has been used for research, or a
simplification of a more polished version.

This type of simplification can take one of several forms. The
most familiar in general geography is when one type of phenomenon or one system is isolated and abstracted from the earth's
surface for the purposes of study. The same abstraction and
isolation takes place in regional geography. Whereas it seems
that no simplification has been made because 'everything' in the
regional landscape receives attention, the region itself is isolated
and abstracted from the rest of the world as though it had no
interconnections at all. In the case of a branch of general
geography, we seem to be aware of, and to accept, the abstraction of the landforms, crops, factories or people from everything else on the earth's surface, much more readily. Even the
study of landforms involves this kind of abstraction. Soils,
vegetation, crops, settlements and man's earthworks are ignored
above, and, except when special attention is directed to
structure, rocks are ignored below.

A second form of simplification, not so readily familiar, but
in common use, is the simplification almost to symbols or
ciphers. For example, in an exposition of settlement patterns,
not only are the relevant phenomena isolated from the relief,

manufacturing, and so on, but also the towns and villages are simplified to dots, the communications to lines, and the fields to patches. The teacher perhaps uses points, lines and patches to represent the elements of the settlement system on his blackboard, map, or overhead projector, and his students visualise, think about, villages, roads and orchards in simplified terms. When a person reads the word 'orchard', does he see, in his imagination, a particular orchard, or just a vague, stylised, model orchard?

Thirdly, even when the phenomena have been abstracted from reality, and modelled by points, lines and patches, it will rarely be the teacher's aim to attempt to cover every aspect of the system in question. Time will be too short, he will leave some things for the students to find out themselves as a necessary process in their education, but many aspects will not be relevant to his theme. Because of this, several simple models or part models may be more useful in teaching than a more elaborate, complete model. The teacher may want to introduce complex ideas in small doses at well separated intervals, he may intend only to examine a few properties and a few well defined factors in the whole system, or he may attempt only one of the many possible types of explanation.

For example, budding geomorphologists have increasingly complex models of the development of a river valley put before them at the fifth form, sixth form, undergraduate and postgraduate stages of their education. All four are necessary simplifications of the system, in so far as it is understood, relevant to the students' powers of comprehension with a given age and store of experience. College students who think they 'did' landforms in the fourth form and do not need to go through it again are infuriatingly unaware of the necessary early simplifications.

The geographers who made themselves notorious by setting out to study *only* the effects of relief, climate, soils and vegetation on man's economic activities provide an extreme example of the type of simplification which isolates only certain features and factors for study and exposition. The very fact that disciplines exist — more, the fact that two or more disciplines study the same phenomena — drives home the point that much exposition simplifies to the stage where it offers only a part explanation. Thus the pedologist studies the rock, climate, vegetation and animal life as factors in the production of a virgin soil. The agronomist is more concerned with cropping, grazing, plowing, draining, irrigation, fertilising and so on as factors affecting the actual soils which farmers use. Economists, when they give attention to farming, have been accused of considering only economic as distinct from physical factors, and at

Commuter zone

High-class residential

Medium-class residential

Low class residential

Wholesale zone

Central business district

(a) Chicago simplified to
    only six functional zones

(b) Model showing all possible features of old-age. Any one real
    example will have only some of these

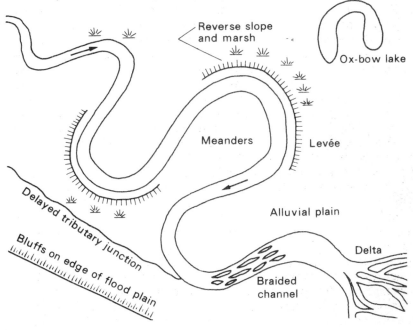

Reverse slope
and marsh

Ox-bow lake

Meanders

Levée

Delayed tributary junction

Alluvial plain

Bluffs on edge of flood plain

Delta

Braided
channel

Fig. 5.1   Simplified and compendium models

the other extreme geographers, particularly at the earlier stages of teaching, have considered what were supposed to be 'geographic' (physical) factors to the exclusion of economic ones.

No discipline, except perhaps regional geography, has claimed or let it be implied that it aimed to study every phenomenon in a place or system. Similarly, none gives a complete explanation. The separate branches of knowledge try to show how things are connected in time *or* how they are arranged in space *or* how they function (Minshull, 1970, pp. 40—1), and nothing more. That so-called geographers have attempted all three types of explanation need not concern us until a later chapter. But it is vital to realise that the model used for teaching and explanation is a simplification and abstraction in this respect as well as in others.

Just as, in teaching, a particular orchard may be simplified to its essentials as a group of fruit trees, so that it represents the essential nature of all orchards, so the simple model can be used to demonstrate the universal features of apparently disparate phenomena. Some students reject Burgess's concentric model of cities, when applied to geography (Jones, 1964, p. 209), as superfluous, claiming that the concentric zones are self-evident. This is possible in the case of two systems being as basically similar as two cities, even if Lincoln and Chicago are several thousand miles apart. The device is used in teaching in another way, when so-called primitive societies, with their simple systems of economy (possibly associated with complex political and social systems) are described and explained first, as an aid to the study and comprehension of more 'advanced' economies later. The work of ethnologists, anthropologists and sociologists is often used by geography teachers in this way (Forde, 1934).

Geography students are now so familiar with the basic, simple, universal model of erosion, transport and deposition that this essential common denominator in phenomena as seemingly disparate as the Amazon basin, a Norwegian glacier, the coast at Skegness and the Australian desert is both accepted and seen as a valuable intellectual concept. In spite of Chisholm's (1970) efforts, one doubts whether basic similarities in widely differing economic phenomena are as familiar to geography students, let alone accepted. But this function of the simple model which demonstrates only the smallest number of essential features and operations in a system is one which can help many pieces of a jigsaw suddenly to fall into place. By sketching the essential outlines (and these must be known from much more detailed models) the teacher can reveal common mechanisms in systems which advance the student's knowledge and promote logical connected thought.

There was a time, less than a quarter of a century ago, when regional geography was dominant and general human geography was in its infancy. At that time, with the exception of those parts of geomorphology and climatology which were labelled 'physical geography', the emphasis was very much on the particular case. Regional geography dealt with particular, unique regions, and still does. But the rather unsystematic branches of economic, urban and population geography tended to deal with specific cases, and made rather feeble generalisations from a mass of half-investigated examples. Since then, three trends have been discernible. First, the trend not only to isolate, say, urban features, from their regions, but also to study them systematically. Second, to concentrate on functions, rather than form and historical development (Taylor, 1949; Mayer and Kohn, 1959), so that laws, rather than generalisations, may be formulated. Third, to use models as the most useful tool in the investigation.

In general geography, then, particularly in research, the specific case is only the stepping stone to the general law. In addition to all the other functions which the model serves, as a simplification it helps to direct attention to the basic elements and functioning of any system, so that the essential similarities between cases can be isolated from local peculiarities. This function of the model does much more than reduce an orchard to a stylised concept, or a town to a vision of little concrete cubes; it represents the essential elements which a river basin, cyclonic storm, farm, factory or town must have, in order to function as such a system at any place and at any time.

This trend in advanced geography, if certain advocates persist in trying to introduce it into schools, is going to conflict with work there. The trend in general geography in schools at present is towards the case study, the investigation of a particular glaciated valley, a particular farm or a unique town. This, in short, is a reaction from the vague generalisations which teachers were making in general geography ten years ago, and which were proving so unsatisfactory because they were generalisations and nothing more. A reaction to the old aims and the old methods has been felt by some teachers and some geographers. But while some teachers have reacted in one direction, to the specific as distinct from the generalisation, some geographers have reacted in the opposite direction, have moved beyond generalisations to the use of models which can embody laws.

8. The use of the model as a compendium

Many textbooks of geomorphology contain block diagrams

showing all the features of glacial erosion and deposition (Holmes, 1965; Wooldridge and Morgan, 1959). After studying and learning about these features, perhaps even redrawing these fascinating diagrams in colour, it must be a bitter disappointment to a student to go, say, to Nant Ffrancon and find that three-quarters of the features do not exist in that area. One can remember being disillusioned with the slick advertising jobs of the diagrams, partly because the drumlins and roches moutonnées did not stick up clearly enough, but mainly because so many of the kames, perched blocks, kettleholes and striations were not there.

Some students realise quicker than others that the model referred to here, comprising a description of the glacial features and an explanation of how they develop, as well as the block diagram, must contain every possible feature of glacial erosion, transport and deposition known to the geomorphologist. Nant Ffrancon may have given the clues which led to the realisation that Britain has been glaciated, but knowledge of the multitude of features which can be formed between the highest peak and the last, lowest deposit of silt on the outwash plain has been gleaned from many examples in many parts of the world. Thus, to be complete, the model must contain every possibility, although it is unlikely that any one glaciated valley on the earth's surface has developed every possible feature. If it had, it would be completely blocked with school parties on field trips.

The model, as a compendium, demonstrates, in the case of glacial action, all the known possible effects of the work of ice, without implying that any one glacier will necessarily produce all of them. The model is still a simplification, because it does such things as isolate the glacial landscape from its surroundings and all man's activities which hide it, demonstrate what one glacier would have done if left undisturbed, as distinct from several glaciations with interruptions, and so on. Yet it is clear that one must think of this type of model as something more complex than any one example to be found in reality, even if it is an abstraction and idealisation in other respects.

Just as any one glaciated landscape may appear to be bare and featureless after the model, the nearest meandering river may lack levées, the cliffs on the coast perhaps have not been honeycombed with caves, and no limestone landscape in Britain is anywhere near as dramatic as the diagrams of Cvijic (Wooldridge and Morgan, 1959, ch. 19). An odd discrepancy here is that schoolchildren and students alike often expect deserts to be flat and featureless, not full of mesas, buttes, pediments, playas, wadis, zeugen and the rest.

The model as a compendium is not confined to geomorphology, although it features in some branches of geography more than others. For example, there are the model synoptic charts which show the deepest possible depression, with the strongest winds indicated by scores of feathered arrows, with perfect warm and cold fronts to gladden the meteorologist's eye, and with a variety of rain, drizzle, sleet, hail and, of course, thunderstorms to cover every eventuality. If such a depression passed over Britain we would not be able to discuss it in a detached, academic way. Again, there is the model of the mixed farm, with every crop and animal possible in Britain, or Von Thünen's (1826) model of farming regions embracing every type of farming possible under the sun.

If the teacher in the field feels frustrated that he cannot point out all the features of river or ice action in one locality, the teacher in the classroom experiences the same difficulty when he refers to examples of industrial location or city growth. The compendium model which contains all the possible factors which may affect the location of an industry has now become very complex indeed. It is so complex and all-embracing, that rarely will all the factors be included in a demonstration or exposition (Hamilton, in *MG*, pp. 370—6). Even when a selection of the industrial location factors has been made for a particular teaching situation, these factors are *all* those considered relevant to a type of industry or a whole region. In the case of one unique factory, or one part of the region, it is probable that only a few of the factors are in operation. Teachers are still searching for the industry which has clearly responded to all factors of industrial location.

Although there are several models of city growth, including the concentric, sector, and multiple nuclei models (Johnson, 1972, ch. 9), each contains all the possible functions of commerce, residence, manufacturing, transport and the rest. When using a particular town to try to demonstrate these zones and functions in the field, the teacher will face the same state of affairs as in the glaciated valley. The compendium models contains much more than it is ever hypothesised that one will find in a particular, real life case.

A variation on this theme will be mentioned here, although its relevance may be clearer after reading the next section. The valley which some students are studying in the field is rather short of erratics and roches moutonnées, but it has a meandering stream above a splendid knickpoint and valley-in-valley feature. The towns which some other students are studying in the map room, by applying the concentric model to examples on ordnance maps, do not show all the features of concentric growth, but do have some clear, wedgeshaped sectors in places.

(a) Burgess
   Concentric model

(b) Hoyt
   Sector model

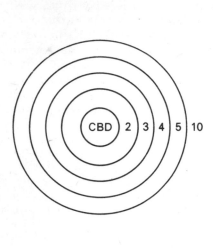

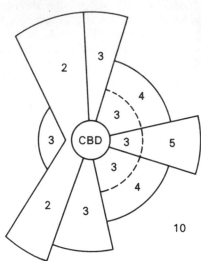

(c) Harris and Ullman
   Multiple nuclei model

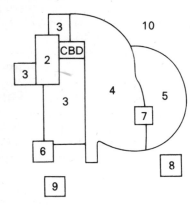

Key to all three models

1 Central business district (CBD)
2 Wholesale and light industry
3 Low class residential
4 Medium class residential
5 High class residential
6 Heavy industry
7 Outlying business district
8 Residential suburb
9 Industrial suburb
10 Commuter zone
   Is one hypothesis correct and
   the others wrong? Does each
   apply to a different type of town?
   Or should a fourth model
   combines features of these three?

Fig. 5.2   Different hypotheses of city structure

Thus one of the greatest difficulties of studying landforms in the field, in addition to the complications caused by several glaciations or several part cycles of river erosion having taken place, is that any one landscape is not just a product of one process; it is a product of glaciation, *and* river action, *and* possibly marine planation at different stages of its development. Similarly, a real town is the product of different processes; sometimes different functions have located on the edge of town as it expanded, at other times industry, commerce and residence have been confined to different sectors. So the students' and teachers' job is sometimes to select a few items from the one model, but much more often to combine items from *two or more* models when analysing a particular phenomenon or region.

9. The use of models as norms and systems of reference

As suggested at the beginning of this chapter, there is a long period of time between the model achieving an acceptable degree of accuracy and its final revision and polishing. During that long period two major aims exist, the one to improve and perfect the model, the other to use the model, however imperfect, for the purpose of studying reality. The two processes aid one another. Most model builders claim that their aims to perfect their models force them to understand the reality which they are trying to model. The more direct purpose of examining real cases on the other hand, inevitably points up weaknesses in the model, and suggests improvements to it.

Here attention will be directed to the use of models, however imperfect, as means to the study of the geography of phenomena on the earth's surface. Mention will have to be made to the question of perfecting models, but I believe that geographers aim to study the earth's surface, and not to construct models for the sake of constructing models.

Three possible objects of the exercise have been implied; the construction of models, the formulation of general geographical laws, and the study of particular cases. Other sections of this book deal with the use of models to help formulate general laws, and the major part of the attention will now be directed to the use of models to examine particular cases. Several authorities are in no doubt that the 'modern' aims of general geography and the use of models are in no way incompatible with the more 'traditional' aim of studying specific cases. Harvey (1969, p. 172) writes, 'the ultimate purpose may be to understand individual cases. This does not mean that a separate theory (or a separate model must be created fo instance', when recommending the use of models to regional geographers.

Lowry (in Berry and Marble, 1968, p. 55) attributes a more traditional aim to model builders when he states 'the model builder . . . is concerned with the application of theories to a concrete case'. However, this is a less direct approach than that adopted by regional geographers; it involves a long detour through general geography and model building. The detour may be so long and so time-consuming that some geographers may doubt whether the new view that they get of a particular region or topic is worth the effort. For example, a traveller, wishing to reach El Dorado, and having no map, must navigate by intuition. He may arrive quickly, but he may never arrive at all. Another group of explorers, with the same aim, spend much time and energy mapping the country. They arrive at El Dorado, much later, but they do arrive, and they have made it possible for anyone else, with the map, to arrive as quickly as possible.

In the same way, the particular, intuitive explanation of a specific case *can* be more effective in the short term — or completely useless. The model builders and users are urging that geographers have the nerve and energy to spend considerable time and effort in establishing the general laws. Many, unlike Harvey, Lowry, and others, do not then see the models and laws necessarily being used for much better regional analysis, but I believe that sufficient geographers will maintain an interest in specific places to be able to make use of models in this way in the future.

In connection with the use of models for analysing specific cases the most frequent references are to descriptive and normative models. For some reason, several authorities use the word 'norm' to mean 'ideal'. In this book the policy has been to keep to the terminology in use, perhaps with some critical comment, in order to avoid confusion; but the words 'norm' and 'normative' are in wide use with clear meanings outside the world of models, and the concept of the model as an ideal has already been introduced, and therefore the words 'norm' and 'ideal' will be used in their correct sense. For example James (in Cohen, 1967) and Taaffe (1970) use the word norm when the models they are describing are in every sense ideals. In addition, other terminology will be introduced in order to label slightly different uses of models.

Most so-called *descriptive models* are in fact norms, but occasionally one gets a hint that a writer is referring to something less general. If this is so, then the descriptive model describes one specific thing, such as the layout and operation of one farm. Such simplified descriptions of individual cases are an essential step in the construction of models, but even a description and explanation of the elements and functioning of one particular system may not qualify for the name model.

In a study of farms, the *normative model* represents the average layout and type of operation, and may not be descriptive of any one farm in particular. This is the correct meaning of the word norm, a typical structure, the average value of observed quantities, not the ideal or the largest. Many farms must be examined, possibly many descriptive models made, before the normative model can be constructed. This will represent the average number, size and layout of the fields, the average proportions and yields of crops and animals, the most common type of working operations, and so on. The norm has become more abstract than the description, in the sense that an average is abstract, but it still refers to how real examples are arranged and operate.

To obtain an *ideal model* of a farm, one would have to proceed in a different manner. Instead of examining every real farm in the field, one would stay at home and pose some questions. If a farmer wants to grow cereals, what is the optimum size and location of a farm? If a farmer has a small hill farm, what will be his most profitable produce? In order to make the maximum profit from farming, and to run a farm in the most efficient way, what is the best size, the best layout of fields, the best combination of crops and animals, the best balance of men and machines, the best way to carry out the work? If these questions can be answered, then the ideal model of a farm can, at least, be put down on paper in the form of a written account, a plan or diagram, some calculations and estimates of costs.

James (in Cohen, 1967) briefly outlines the procedure for making and using an ideal model, saying that one describes what the spatial systems should be if the goals were X and the constraints Y and Z. Then the solution of the equation provides the norm (*sic*) against which actual conditions can be measured. Possibly in a sense different from that proposed by Harvey this is an *a priori* model. Taaffe (1970) points out that the construction of such ideal models is a departure for the geographer from asking why things are where they are and how they do work, to asking where they ought to be and how they could work better. So many geographers in the past have blithely assumed that the way man has arranged himself on the earth's surface is the best of all possible ways, that this new departure is infinitely valuable if all it does is make the geographer aware that something better is possible, and that man has not always had the complete information to enable him to reach the perfect solution.

The concept of the ideal model thus has several implications. The ideal may be something for the farmer to aim at; or it may be something for the planner to try to bring about. The ideal can be something against which the geographer compares reality, both to understand reality and to realise that man has much to learn about arranging himself on and using the earth's surface more effectively. There is the danger that the use of the ideal might become derogatory and disheartening, if it were always assumed that reality had to match up to the ideal. So it may be necessary at times to stress that the ideal model is being used only as a frame of reference, as a measuring stick and not as an exhortation.

The descriptive, normative and idealistic models are used as frames of reference, but in certain circumstances they smack too much of reality or of an impossible standard. In such cases an *abstract model* will be more useful as a frame of reference. For example, the concepts of random, regular and clustered distributions are used as standards against which to compare the real distributions of such things as rural settlement and non-residential functions in cities. Such abstract models as the random distribution and the nesting of hexagons are not norms which have emerged from empirical studies, and they are certainly not ideals at which anyone is aiming. They do, however, serve extremely valuable functions in the analysis of spatial patterns on the earth's surface.

Haggett uses several examples of such abstract models in *Models in Geography* (p. 609) to analyse road networks and stream patterns, and in *Locational Analysis in Human Geography* (p. 89) to analyse settlement and other point phenomena. Bunge (1962) makes reference to similar uses, although his main interest is in the geographical theory which might be developed through the application of abstract concepts.

All four of the models outlined above share certain uses. They are all used to study particular cases, to help discard the irrelevant, and to show that any real landscape is made up of a combination of many phenomena. The main use of the descriptive model, to continue the farming example, is to lay bare the spatial layout of the farm and its economic operation. In this type of study the material of field boundaries, the architecture of the farmhouse, even the nationality of the farmers, are discarded as irrelevant. This stripping to essentials, to reveal the basic layout and functioning, can proceed from farm, to region, to continent, to a world scale, as well as from farm to farm. In a most convincing argument, which is a superb example of the modern aim of geography, as well as the use of modern methods, Chisholm (1966) lays bare essentially similar mechanisms within one farm, within one continent, and at the level of world trade in agricultural produce. He demonstrates how an analytical tool can bring order out of seeming chaos, and enable us to comprehend several diverse phenomena.

Once a geographer has defined his aims this will also help to define what is irrelevant, and make more certain the construction of the descriptive model. However, there are many instances when, for a long time, the geographer cannot be sure what is and what is not irrelevant to his purpose. He may aim to study the natural vegetation of forests, but, once in the field, the phenomenon in which he is interested is so closely affected by everything else in the landscape that often it is impossible for him to decide what is a feature of the vegetation, what is a distortion due to the effect of other phenomena, or man's actions, and what is just a random or purely local variation.

Model procedures, mathematical and cartographic techniques, are being developed to help with this problem. Details are to be found in books explaining quantitative techniques in geography (e.g., Gregory, 1973; Cole and King, 1968) and one example will suffice. Board (*MG*, p. 716), describing trend surface mapping, shows that it both cuts out 'unexplained random variations' and is of great value in 'sifting the regional from the local component in spatial patterns'. Comparisons between topographic maps, trend surface maps, and the maps produced intuitively by the traditional 'eyeballing' method show that the latter are too subjective and misleading. Standard procedures for removing irrelevant detail in maps ensure that there are precise criteria for leaving in or removing information, and so ensure that the same result can be produced by different geographers, or the same geographer at different times.

The general geographer may have an irresistible urge to tidy up his data, both to remove it from the untidy earth's surface and to remove the local variations in which he is not interested. However, by its nature, the descriptive model will have to be an 'untidy' thing if the system it represents is in any way untidy. We may like to think that man is completely rational, and certainly many economic geographers show that they would prefer man to be influenced only by economic motives and factors. But even in less extreme cases than the support of sacred cattle in India, or the use of cattle as currency and objects of admiration rather than sources of food in parts of Africa, man is 'irrational' and 'untidy' in the particular meanings of the words used here. A descriptive model of some

(a) Descriptive model. Location of steelworks,Whyalla South Australia

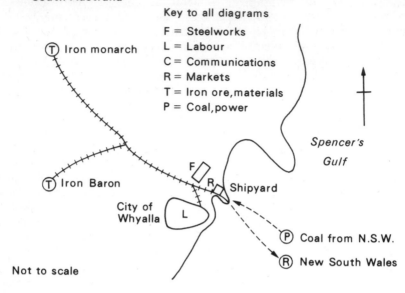

Key to all diagrams

F = Steelworks
L = Labour
C = Communications
R = Markets
T = Iron ore,materials
P = Coal,power

Iron monarch

Spencer's Gulf

Iron Baron

F

R   Shipyard

City of Whyalla   L

P Coal from N.S.W.

R New South Wales

Not to scale

(b) Normative model   **(c) Ideal model**

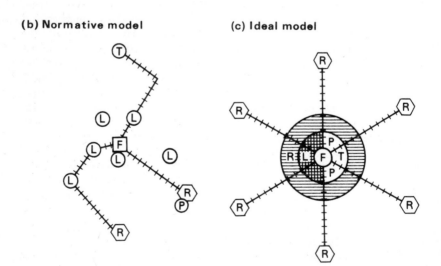

Fig. 5.3   (a) The layout of one particular system. (b) The average layout with materials and markets separate and labour dispersed. (c) All factors of production at one place and many markets

farming will represent a most inefficient scattering of plots and strips of land over a wide area, and unintensive operation. Another model will represent factories operating in a region long after the original location factors have ceased to have any effect.

A normative or an ideal model can only be of use in analysing particular cases if it already bears some relation to the particular. Most of the ideal models, and many of the normative models, are so far removed from the actual conditions which obtain on the earth's surface that many critics are rightly suspicious of them. The neat normative model which shows the majority of people living in the fertile lowlands, avoiding both extremes of temperature and extremes of rainfall, does not apply in parts of tropical Africa and Latin America. If one wants to begin to analyse and explain the seemingly haphazard distribution of population in tropical Africa, then a starting point much nearer to the real state of affairs is essential. In such cases, a particular descriptive model, however crude, may be the necessary first step, however attractive the more elaborate models are, in analysing the real situation.

The normative model is usually considered to be much more useful than the descriptive model in the analysis of particular cases, but there remains the problem of achieving a balance between the descriptive and the norm. A balance must be achieved between the model which is so specific that it will apply only to one case, and a model which will apply to every case, but is so general and vague as to be worthless. The question arises whether a normative model should be applicable to all cases of the phenomenon being studied. Some workers suggest a number of normative models and others suggest a model with variable components; McCarty and Lindberg (1966, pp. 9—10) seriously consider both possibilities as practical solutions to a perennial problem. Objections are raised to both solutions on theoretical grounds. Haggett (*MG*, p. 26) objects to a number of models, especially when applicable to different regions, on the grounds that a model must be universal. Guelke (1971) objects to models with variable components on the grounds that such a model can never be tested rigorously, and is further varied as an excuse for it having 'seemed' to fail a test.

So many techniques are now offered to the geographer to help him establish the norm, the average, the mean, that attention is directed away from the particular, and certainly away from the abnormal and unusual. The normative model of a system, such as a farm or a farming region, is much more complex than a numerical average but it bears two characteristics in common. First, the norm, like the mean, does not necessarily represent one particular item from the total population.

Second, the norm, like the mean, can be used either to represent the total population, or as a reference point against which to measure a particular individual. Some model users are more interested to use the norms as representatives, to help in the formation of general laws about the operation of the phenomenon in general, but the normative model is equally useful as an analytical tool and a yardstick, to measure how different from the norm a specific case is.

The geographer, investigating a unique farming region, may suspect that there is something special about the farms; but he is completely unable to say what is special until he has a norm as a standard. Analysing the farms by means of the normative model, significant differences such as size, number of fields, crop rotations, cooperative practices will become clear. Moreover, they can be claimed as significant differences because the norm exists. Then it is worthwhile going on to look for explanations of this particular regional variation. Many geographers in the past have devoted much time and energy trying to explain regional variations which, by more stringent tests, would have proved to be not significant, not worth attention.

A different use of the normative model lies in examining the influence of different factors on the norm, rather than variations from the norm itself. Urban geographers have long been interested in the effect of relief on town growth. In particular case after particular case they have claimed to study how the site of each town has aided the growth in one direction and restricted the growth in others, to give each town a unique plan, depending on the local relief. Most of this is doubtful. To make any kind of scientific statement about the effect of relief on the growth of a town, one must have either an identical town, growing as a 'control' on a completely flat surface, or one must have a norm, constructed from such a large number of towns that individual features resulting from special growth and unique relief have been completely cancelled out. In geography a control for an experiment is out of the question, so for research work which is to have any validity at all the normative model is of supreme importance as a substitute for a control.

The factors which interfered with the norm do not have to be physical. In the study of farms one would be less likely to come across cases of high relief distorting arable farms, than, say, cases of the economic influence of cities producing changes in the normal farm. Frequently, geographical descriptions refer to the growth of cities out into farming areas having the effect of causing a rise in land values which forces the farmers to go in for much more intensive activities, such as market gardening; or the demand from the city causing more and more farmers to change to dairy farming. It sounds plausible enough, but it would carry more weight, and be of more help in understanding the process, if we were told just what the original average farm was, and what proportions of the total population were already market gardens and dairy farms. Again the norm must be established first.

In most books on quantitative techniques the writers take care at the start to expound the mean, the median and the mode. It is important to understand that these are different norms, referring to different characteristics of a population, worked out in different ways. Now, while normative models may not be worked out in such precisely defined different ways, there is much to suggest that two or more norms are necessary for one type of phenomenon. For example, if it is understood and agreed that the study of the effect of site on town growth demands a norm of town growth, will the norm which applies in Europe apply in West Africa, in India, in China, in Ireland as well as in southern Italy?

Geographers believe, with some confidence, that models of physical systems, river basins, land and sea breezes, and soil profiles apply equally well all over the world. But are the economic functions, the residential behaviour, the social aims of every culture, every nationality, every political and economic system anywhere on the globe so much the same that the geographer can create one universal norm? It is much more probable that one normative model, which could include any economic, social or political factor which had an effect on town growth in any type of urban society would be so general, so vague, that its value as a norm to test a particular case would be lost. It would be about as useful as an elastic tape-measure.

A possible way out of this dilemma is the construction of an ideal model which can be used instead of the norm to evaluate particular cases. If it is impossible to agree on one norm, and theoretically undesirable to have two or more, then surely the ideal can be worked out, even if it has never been attained. Many architects have planned the ideal city, and Christaller has proposed the ideal layout for a large number of towns and cities, so an ideal model of a much simpler system should be easy to make.

Given a farming community in which each family owns land and a house; a place of worship, a general store, a school, a hospital and a market depot are essential items, and the British geographer sets out to construct the ideal spatial layout which will be his system of reference for the spatial arrangement of any farming community in the world. Parcel each man's land together, put each house in the middle of the plot, group the public buildings together in the centre of the total area and the job is done. The ideal answer is obvious to anyone with any sense.

The geographer is convinced of his genius when the farmers are satisfied. Not only has he put the houses on the land, he has placed them centrally so that the farmer has the minimum distance to walk to any field. The children are not so convinced, they have to walk to school every day. The wives are furious, not only is the shop so far, but all their friends are scattered, and the church, the store, the school, hospital and warehouse do not provide much attraction for the teenagers on Saturday night. Then some farmers away on the edge of the community, who don't know what's good for them, start to complain about the distance to market, and that while a British system might be ideal for some, being dispersed makes it difficult to get the communal planting and harvesting done.

Perhaps a village is the ideal answer after all. But if the geographer constructs an ideal model of a village to apply to the whole world, just as many snags will arise for different reasons. The children, teenagers and wives may be content enough, but the farmers' complaints about access, distance to work, unnecessary movements of machinery, cost of moving fertilisers and produce will be multiplied. To suggest two ideal models, one nucleated, one dispersed, defeats the object of the exercise, of course.

The illustration was developed at some length to emphasise that what is ideal for one community is far from ideal for another. British and American geographers, together with some Europeans, have been far too presumptuous in setting up ideal models which in fact are only European or American ideals. The arrogance is greater the more it is confidently assumed that these ideal spatial arrangements will be equally and absolutely ideal to any peoples on the earth's surface. One could wish for a flourishing Dayak school of geography, or a well developed set of Red Indian theoretical models to counteract the excesses of Chicago and Lund.

On a more charitable and realistic note, it must be admitted that the so-called ideal model can be used as a frame of reference, even if it is a 'Western' ideal, for the purposes of analysis. The geographer can take one norm, one biased ideal, of the layout of a farming region or of a city, and use it to analyse the actual layout of a particular region or a real city. At times there will be major discrepancies between the model and the reality. But providing that the geographer directs his attention to the reality, and not to either the discrepancies or the model, then this procedure is perfectly valid. Perhaps the most important point is for the geographer to make absolutely clear that while he is using a so-called ideal model, there is no implication either that this *must* be the ideal of the people in the region in question, or that they have 'failed' to reach the ideal. Many geographers use a temporary model in this way, setting up some kind of norm or ideal purely for the purpose of one investigation of one particular topic. Once the work is complete, the model has served its purpose as a frame of reference or as an analytical tool, and is scrapped. In such an investigation there is no intention of using the findings to further polish and perfect the model.

Frequent reference has been made to variable factors within a model, and to variations from the norm, but in both cases the implication has been that these are variations *within* the system, or within the model of the system. For example, one has in mind the variation of the economic factors affecting farming, or the variations of farms from the normative model. When models are being used to analyse specific cases a completely different set of variables obtrudes. These are elements and forces from *outside* the system, which affect real life examples of the phenomenon under study in a way perhaps not allowed for in the model. A model of the distribution of central places will allow for different spacing of the towns in different regions, say, according to the density of population. The density of population would be considered as part of the settlement system, and a mathematical expression could link the spacing of towns to the population density. In contrast, the same model will not allow for the effect of such things as relief on the spacing of towns. Relief and other physical elements of the environment are usually considered to be outside the settlement system and so are not allowed for in the model. Consequently, when a normative model representing the spacing of towns on a perfectly flat, perfectly uniform plain, is used to help analyse the spacing of real towns in a real region, the obtrusive factors from outside the system can cause considerable confusion.

Some models of farming systems exclude the effects of relief, climate and soil, just as others exclude the effects of tariffs and political policies (Henshall, *MG*, ch. 11). Until Pred's (1964) work, models of industrial location ignored the effects of central places. Above all, the central place model is concerned only with economic factors, and excludes the effects of relief, climate, soil, and vegetation, although most geographers would acknowledge, at least, that man is influenced by relief in siting his towns. Probably because man, in his turn, has such a limited effect on the natural features, it is never so obvious that models of landforms, climate, soils and vegetation usually exclude the effects of man. Man has not done much to change the relief of the earth's surface, but the geomorphologist still ignores dams, groynes, dredging, sea-walls and the like. Seeding clouds and exploding hydrogen bombs have not changed the climate yet, but man has transformed most of the soil and 'wild' vegetation on the face of the earth. Just as some geographers persist in trying to re-create the original natural vegetation, others persist in creating settlement patterns on plains even flatter than Lincolnshire, without the slightest trace of a 1 metre variation in relief.

So, to the physical geographer, man has interfered with something he would like to study in its original form, for its own sake. To some human geographers, the real surface of the earth simply interrupts the systems which man has tried to create, or is a distraction to be ignored as much as possible. The geographer interested in the central place model, unwilling to be distracted by the effects of high relief, must be a very different fellow from the geographer who once aimed to study how man adapts himself to his physical environment; or from the geographer who still studies man—land relationships (Eyre and Jones, 1966).

One type of investigation, then, of a particular case, needs to ignore the effects of relief. The investigation of the number, size and spacing of settlements in eastern England aims to compare the actual with the normative model. Where the distribution is not as expected in the fens, wolds and moors, the normative model is defended, and it is argued that the central place model presupposes a flat plain; where there is high relief the model is not expected to apply. So the interest in this type of investigation is not where the relief causes a deviation from the norm, but where there are deviations from the norm on the more level parts of eastern England, where the norm *is* expected. Critics of the central place model believe they have successfully rejected it when they can show how badly it fits the situation in the Pennines. Supporters of the model insist that this does not weaken their case, because the model was never intended to fit such situations.

For the supporters of the central place model, a much more serious situation arises when, in level country, the size and spacing of central places is not as the normative model predicts. When, in Lincolnshire, Kirton Lindsey is only one-third of the expected size, Grantham is 5 miles too far west, and Louth offers far more retail services than the size of its hinterland seems to demand, important decisions have to be made. When a model of an incomplete theory is being used to analyse a particular real case, and discrepancies between model and reality are found, three possibilities exist. First, that the discrepancies are not significant, and just arise from elaborate details which have been simplified out of the model. Concern over such things can waste time, especially when one geographer's model is being used by someone else, who is so far removed from the original work that he does not realise that such details or 'noise' have been simplified out of the model.

Second, the differences in reality from the model may be important, but purely local features. In the example above, Kirton Lindsey and Grantham certainly do need special explanation, and the normative model has served its purpose in helping to analyse a special area, and to isolate those particular features neither accounted for in the model nor too trivial to ignore. Third, the differences in reality may be so numerous, or of a kind which have been met in other investigations, that a modification in the model is indicated. If the majority of towns in East Anglia are further apart and smaller than in the normative model, there may be a particular reason for this in the region, but there may also be a basic fault in the model. Similarly, if two or three investigations reveal that towns in Britain have more attributes and variates than listed in the normative model, then the model will have to be modified to represent the normal state of affairs.

When models are being compared with reality, this is perhaps the most important question which can arise. Does a discrepancy between the model and reality demand a special explanation of this one real case, or does it demand a change in the model? The only safe answer which can be given to this question is to say that when this situation arises, then many more real cases must be examined. If many more discrepancies and exceptions are found, then clearly the model must be modified; if the cases and the model match one another, then, equally clearly, the instance which posed the question demands special attention.

Once the question has arisen, it is vital to do one thing at a time. Either use the model to analyse real cases, or use real cases to test the model. Any student can get into serious difficulties either by setting out with the hope that he can do both at once, or by losing sight of his single aim and, say, setting out to examine real cases and ending up by changing the model. In any single investigation, either the model or reality must be taken as a fixed point for that investigation.

It is precisely in this third type of situation that one can lose sight of the object of the exercise. If a piece of work which set out to examine the settlement of eastern England ends up by changing the central place model, critics will have grounds for claiming that the geographer does not know what he is doing.

101

The criticism is most strong when the details of a particular piece of work give the impression, usually completely wrongly, that the investigator has changed his yardstick halfway through his measurements, and in effect, has distorted his findings. Many authorities refer to this type of situation, most of them accepting it in passing. McCarty and Lindberg (1966), however, welcome it as something which can be turned to advantage; referring to a graphical model of farming, they write, 'once we are in possession of such a space model it is quite possible to modify it to make it more applicable to the area we are investigating' (pp. 60—3). There is no doubt in their minds either that they are investigating particular cases, or that the model can be modified to fit those cases. Guelke (1971), on the other hand, is most sceptical of this changing horses in midstream, and considers it to be a serious weakness of the use of models in geography.

This is a realistic state of affairs. There are few really well tested models in geography as yet, and there is no question of waiting before what models do exist are put to use. Moreover, it is too much to demand that a geographer either uses a model to examine cases without modifying the model, or goes about testing the model without taking an interest in specific cases. All these aspects of the work are inextricably bound up in each other. All one could hope for is that during a particular investigation, when the geographer's experiences indicate at last that some modification of the work is necessary, the paper which he finally publishes makes absolutely clear the change of direction necessitated during the work.

One authority, at least, envisages the possibility of the geographer aiming to study the effects of relief on the spatial arrangements of a system. It was argued earlier that an urban geographer who wanted to be able to say how the relief of a particular site had affected the growth of a city must have a norm of city growth on a flat plain. Nystuen (in Berry and Marble, 1968, ch. 3) argues that in order to make sense of the arrangement of some phenomena in areas of high relief, it may be necessary for the geographer to exclude the relief first, get an idea of the ideal layout, and so be in a position to appreciate the effects of relief later. The illustration Nystuen gives is of pupils being led out of a mosque into rugged terrain for a lecture. Their spatial arrangement in the rugged area outside makes no sense at all, until we see how the pupils are in the habit of arranging themselves in a ritual pattern around the lecturer on the perfectly flat floor inside the mosque. Once we understand this pattern, we can understand how the pupils have attempted to achieve the same pattern on the very difficult relief outside. To an Englishman, used to the lanes and roads which take easy routes, winding about up the scarp in Lincoln, or up the hillsides in Halifax and Huddersfield, the streets straight up the hillsides in San Francisco seem decidedly senseless, until one realises that they result from the imposition of the rectangular land division, appropriate to the prairies and plains, to the steep hills round the bay.

In such a case, of course, there is no question of modifying the model. Models of the spatial arrangements of farms, mines, factories, towns, communications and population can be constructed as they might be in isotropic conditions — in conditions of flat land, uniform rock, soil and climate, so that the effects of one or more of these factors can be investigated. A different type of danger here is that although the effects of relief may then appear to be self-evident, even this is not certain, and variations of the spatial arrangement due to differences in soil or climate may be much more difficult to isolate and establish.

An interest in the way physical conditions do modify ideal layouts was much more noticeable in geography before models were used in any systematic way. This interest is still essential in agricultural geography, but the development of models in other branches of human geography, for the time being, at least, is leading to an interest in systems for their own sake, not as they exist on a real portion of the earth's surface, modified by all the other systems occupying the same area. In Nystuen's example, to understand how rocks, relief and climate have affected the spatial arrangements of a group of people, first, the geographer must have an ideal model of those arrangements which can be modified by other factors. This would have been appropriate to the work of determinists, if the use of models had been as explicit in the past, or if there was much interest in determinism today. In order to be able to say that physical conditions have determined, or even affected, a people's way of life, the geographer must have a model either of their way of life under another set of conditions, or of their way of life under isotropic conditions. Similarly, in order to be able to claim that a region is a medal struck in the likeness of the people, one must have, at least, a model of their activities in another region. Although models are unlikely to be used to investigate problems of determinism, such suggestions help to emphasise both that these new methods can offer new solutions to perennial problems, and that some of the problems investigated in geography will demand very sophisticated techniques if they are ever to be solved.

When a model is being used to analyse a system in a

particular landscape, it soon becomes clear that one model, of whatever system, will help to explain only a fraction of the whole. For example, in eastern England, the central place model certainly applies to the vast majority of towns and cities. But that model specifically excludes break-of-bulk points and resource-based points. In other words, it excludes manufacturing towns such as Scunthorpe, ports such as Hull, and resorts such as Hunstanton. Other models, of the location of industry and of other economic activities are necessary to help analyse these. In the same way, it is necessary to refer to models of geology, of river erosion and of glacial deposition, in order even to begin to analyse the landforms. This is where so many students fail to make the change from geomorphology to the geography of landforms. The geomorphologists keep the systems of river action, glacial action and the like, artificially separate. Unfortunately for some students, such systems are not separate on the surface of the earth. The landforms of Lincolnshire cannot be understood by reference to the chapter on river erosion in a textbook of geomorphology; one needs to refer also to chapters on glaciation, coastal erosion and deposition, the influence of structure, and limestone landforms. This is not to advocate a return to the geographical description which is a hodge-podge of scarps, moraines, cliffs and fens. The modern geographical approach can be analytical, but each of the separate branches of geography provide the student with only one of the *many* models which he will need to apply in a particular area.

The particular interest in the landscape will determine how many different models will be needed in its analysis. If the interest is confined to the landforms of a region, then it may be necessary to compare the real landscape with models of ideal or normative landscapes produced by river, ice and sea action. This analysis of complex reality by comparison with a series of separate idealised models may enable one to comprehend how the particular landscape of, say, the Midlands or Wales is a combination of partially developed landforms produced at different times by different processes. The models separate them neatly, but the landscape is the resultant of all their interrelationships.

Similarly, the analysis of an ecosystem may demand the use of models of landforms, climate, microclimate, soils and vegetation associations for one to understand how even a very small area of the earth's surface contains a complicated dynamic interaction of the systems isolated and simplified for study in the geography room. The separate models which explain the location of farming, mining and manufacturing again have to be brought together if the aim is to analyse and understand the complete economy of a particular landscape or region. Even more, if one's interest is in the whole of the human geography to be seen in a landscape, the economic models will have to be augmented by models of settlement, communications and population.

If one's interest is confined to only certain aspects of the humanised or cultural landscape, again if the interest is in the real, complex landscape, then it will be necessary to measure it against several models in order to analyse it. For example, the settlement pattern may be a mixture of Christaller's (1933) K3, K4, and K7 systems, and the transport networks may be both branching and connected circuits. In particular, experience suggests that none of the towns will have just a concentric plan, or just a sector plan as the present models hypothesise. Different processes, operating at different times in the towns, just as different processes of river and ice action have operated at different times on the landforms, have produced very complex situations in real life. Therefore it seems that several models, of different types, will be needed in the analysis of just one landscape.

Finally, a few of the more obvious dangers of using models for analysis must be summarised. When they first use models in this way, many students deviate to testing the model rather than analysing the landscape. Instead of concluding by making some statement about the geography of a selected topic in a particular place, such as the farming in the east Midlands, they end up by defending or criticising the model. This is the most depressing aspect of the use of models in geography, to be found among research workers as well as new students, that the model attracts attention to such an extent that the geography of real places is ignored.

A variation on this theme is to refuse to see discrepancies from the model either as local features which deserve special explanation or as significant phenomena which necessitate modification of the model.

Lack of interest in the landscape results in special cases being neglected, while admiration for the model prevents any suggestion that it is less than perfect. In such situations the deplorable course of action is to 'explain away' the discrepancies, resulting in no progress, neither improvement of the model nor a better understanding of the earth's surface. For example, Cole and King (1968, pp. 522—3) make it clear that the study of a real situation is of secondary importance, while, when model and reality do not fit, it is reality which has been misunderstood, not the model which is wrong.

To the dangers of ending up with the model, and making reality fit the model, Guelke (1971) adds the danger of studying only the discrepancies between the model and reality, perhaps the greatest danger of all. When the normative model and a real case do not coincide exactly, as is more than likely, then so much attention may well be given either to explaining special local circumstances, or to modifying one small aspect of the model, that the major part of the model, and the bulk of the phenomena in the real system which do coincide satisfactorily are ignored. In that case the student has ignored exactly what he set out to describe and explain.

6
Analogies, model building, black box and noise

10. Analogies

Chorley and Haggett (*MG*, p. 23) state, 'because models are different from the real world they are analogies'. Cole and King (1968, p. 497) state:

> the essential difference between models and analogies is that in the former, either smaller or larger replicas of the original or symbols are used to stand in for the prototype features. In the latter the features compared are quite dissimilar in nearly all respects except those that constitute the positive analogy.

Harvey (1969, p. 147) writes of 'the failure to differentiate between models and analogies', while several classifications include analogues as one type of model.

From these authorities on models in geography we have every logical possibility: that all models are analogies; that no models are analogies; and that some models are analogies. Therefore the reader will not be surprised if no definite final statement is made about the matter at the end of this section. The aim is to discuss some of the conflicting ideas, and to prepare the student for the misunderstandings and confusion which abound in the literature on models. It is tempting to take the easy way out, with Chorley (in Berry and Marble, 1968, ch. 4), and use 'analog' as a synonym for 'model', but Harvey is concerned that models must be kept separate from analogies, and that confusion of the two will lead to serious error.

It is very difficult to accept the reason Chorley and Haggett give for all models being analogies. If this were sufficient reason, then *any* representation of a thing would be an analogy, and the word would have such a wide meaning that it would be virtually useless. Chorley, in a separate article, however, makes it understandable how such a confusion of terms can come about. Writing of geography and analogue theory, he refers, as do many geomorphologists and economic geographers, to the model as necessarily a mathematical model. Now mathematical models are generally for use either in digital or in analogue computers, and it becomes common to think, first, of a model as an analogue computer model, then as an analogue model, and finally as an analogue or analogy. Many other writers clearly use 'analogue model' and 'analogy' as interchangeable terms.

In the strict sense, one uses as an analogy something which exists already, but Chorley (in Berry and Marble) writes of 'model building, or analogue theory, in geography'. He begins his article by referring to analogies as such, but soon combines them with models, model-building becomes analogy-making, and it becomes clear that Chorley can conceive an analogy as something one makes. He insists that the manufactured model/ analogy must be something more familiar than the system which it represents. As this is the original idea behind the use of analogy, to explain the working of the unfamiliar by reference to the working of the familiar, one must agree that Chorley has the essence of analogy. But enough has been written about models in earlier chapters for it to be clear that there are many useful models of which the working is not at once familiar and easy to understand. The central place model, one of the most elegant and effective in geography, is at first unfamiliar and difficult to grasp, yet in the end most valuable in explaining many aspects of settlement geography. If only in this matter of familiarity, then, one must insist on a difference between model and analogy.

It is easier to accept the idea that an analogy may be a manufactured thing, rather than something which already exists and is just selected because of its similarities with the phenomenon being studied. If the phenomenon being studied is the movement and migration of population, then a natural analogy for this would be the movement and migration of animals. The animal system exists in its own right, is of independent origin, but has functional similarities with a human system. A geographer studying movements of people, however, might perceive some similar type of movement and flow in an electrical circuit, and find this a useful analogy in certain respects. Originally, electrical circuits were manufactured for completely different purposes; again they have an independent origin but one or two functional similarities, but eventually a specific circuit is built to represent the flow of people. Whether this is a manufactured analogy, or whether it has now become a specific experimental model is open to debate.

In the same study of population movement there is the possibility that the geographer will make his own model to represent the migration. This will take the form of words and statistics to represent the population elements of the system, maps and equations to represent the functioning, and an hypothetical explanation. The model will rely on symbolic (verbal, graphic and mathematical) representation of the population system, and not on completely different material things such as animals or electrical circuits. Another possibility is that the geographer will borrow a model made by someone else, for a different purpose, but which he thinks will help in his study of population movements. Such a model could be the gravity model, which compares the forces of attraction of two towns with the gravitational forces of two stars or planets in space. Because the influence of a town of a given size is observed to decrease the further away from the town, just as the gravitational attraction of a planet of a given size decreases with distance away from the planet, the model of gravitational attraction is believed to be equally applicable as a model of urban attraction. One model, in this case the mathematical model

$$M = \frac{P_i\, P_j}{d ij^2}$$

is used to represent two completely different systems.

Moreover, in addition to the fact that a model made to represent one system can be used to represent another, is the fact that the explanation of the operation of one system can be used as an analogy to help explain the operation of the other. With fifteen Apollo space flights completed, it is believed that readers will be familiar with the process of a rocket being shot towards the moon until it passes the 'break point', enters the moon's field of gravity, and falls to the moon. On the return journey the rocket is shot up from the moon's surface until it passes the break point and falls back to earth. Similarly, a break point is imagined between two towns. On one side of it, people will move toward town A, and do their shopping there, on the other side they will move toward town B, and do their shopping there. Just as the break point is much nearer the tiny moon than the large earth, the break point is much nearer the smaller town, and a formula for finding the break point is (Ambrose, 1969, p. 163):

$$Db = \frac{Dab}{1 + \dfrac{Pa}{Pb}}$$

where Db = break point in miles from b

Dab = distance between towns a and b

Pa = population of town a

Pb = population of town b

Thus we see the possibility of four ways of representing at least some aspects of a system:

1. by a model made for that system;
2. by a model made for another system, and borrowed;
3. by an analogy, using a manufactured or artificial phenomenon;
4. by an analogy, using a natural phenomenon.

Cole and King also mention greater familiarity with the analogy as one of its useful features, but they show that however familiar the analogy may be, it also has important drawbacks. In order to study the traffic flow in a city centre, one could construct a model of the traffic flow or use a set of electrical circuits as an analogy. When comparing the model with the real thing, one is comparing like with like, but the major drawback of comparing the analogy with the real thing is that one is attempting to compare two things different in kind. Not only are they different in substance and operation, but also the analogy will contain features and subsystems which just do not exist in the real thing. However crude a model, the builder aims to represent only the features of the system. However good the analogy, it will contain some features which have no counterpart in the system of traffic.

Sometimes, of course, the familiarity of an analogy does help to get ideas across. W. M. Davis modelled the development of a river system, describing and explaining how the features of the river and the valley developed and grew in time, words, maps, examples and diagrams all playing their part. But when Davis used the analogy of human growth, suggesting that the huge meandering river on a wide plain was once a tiny stream in a valley with interlocking spurs, just as the old man was once a baby, many ideas clicked into place. The analogy of youth, maturity and old age, although scorned in some quarters now, helped many people to see a river basin as a dynamic system. Youth, maturity and old age is not a model of a river system; it is an analogy for its growth, and many features of human growth have no counterpart in the river system at all.

The main problem with the use of analogies as something more familiar than the real thing or than the model is that what is familiar to one person is by no means familiar to another. Some of the analogies used by Haggett are decidedly obscure, for example, many geography students are more likely to be puzzled by electrical circuits than they are by traffic flow diagrams. A difficult idea to get across in teaching is how a river can cut its head backwards, away from the sea, when the water is flowing fowards towards the sea. A circular saw was chosen as an analogy, the teeth moving seawards but the whole saw biting into the land. The boys got the point at once; the girls were even more confused.

McCarty and Lindberg (1966, p. 8) use the word analogue in a slightly different way again. Making reference to the fact that geographers can neither experiment on the real phenomena in the field, nor bring the phenomena into the laboratory for experiment, they write of carrying out experiments on an analogy, or analogous system, which can be brought in to the laboratory. While they use the words analogue and analogous it soon becomes clear that they mean mathematical and graphical models, and simulation procedures. In this respect McCarty and Lindberg parallel Chorley and Haggett in using such words as analogy to refer to devices which are more widely called models.

The longest and most serious argument about the difference between models and analogies is reviewed by Harvey (1969, ch. 10). His discussion is difficult to follow, and it is probable that part of the confusion stems from a failure to distinguish between what have been called parts I, II and III of the model in this book. The authorities Harvey quotes are at cross purposes because one refers to the structure of the system while the other refers to functioning of the system *in the same words*. The argument between Braithwaite and Achinstein, reviewed by Harvey, stems from the original idea that an analogy is used only to parallel the function of the system. Achinstein can see no good reason why the analogy should not also parallel the structure of the system or even the physical material of which it is made. Braithwaite's objection is that as the analogy is something different from the system in the first place, the more one tries to see parallels in every aspect of system and analogy, the more one is distorting one's concept of the original system to fit one's better understanding of the analogy. Long ago, Hartshorne criticised this overuse of the analogy when he objected to regional systems being compared to living organisms. In the extreme cases regions were being described as if they were organisms, animals, which needed to grow and reproduce. So much of the analogy had been read into the system that, far from understanding the system, certain geographers had distorted it out of all recognition and really did think they understood regions because they could understand some aspects of animal physiology. The worst excess was when geopolitical geographers defended Hitler's conquests on the grounds that the

German state needed room to grow and develop like an organism (Hartshorne, 1939).

Cole and King, and Harvey, draw attention to the concepts of positive, negative and neutral analogy which try to guard against this danger, although they have slightly different definitions of the terms.

> Those aspects (of object and analogy) that differ are called the negative analogy and the other features that do not concern the problem under consideration or which are non-comparable or irrelevant are the neutral analogy and can be ignored. It is the positive analogy which is of significance in using the method of analogy . . . for research purposes (Cole and King, 1968, p. 498).

Harvey (1969, p. 150) refers only to the physical properties of the object and the analogy as the same (positive) or different (negative), and he gives a different definition of neutral analogy which exists when the relationship is not yet established. To say that a relationship has not yet been established is a different thing from saying that the relationship is non-comparable or irrelevant. Having given the warning that parts of the analogy are at best neutral and at worst negative, these authorities give no guidance on how to recognise and reject neutral and negative analogy.

Thus the model contains, or should contain, only elements and functions believed to be in the real system. They have been selected because the model is a model of that system. An analogy, which has been selected for having some similarity with the system, also contains elements and functions which parallel nothing in the system at all. The analogy may be a different natural system, as cliff slumping has been used as an analogy for ice rotation in a corrie; it may be another manmade system, or it may just possibly be a model of another system made for another purpose. The analogy is therefore different from the model and, at the same time, much more dangerous to use because it does contain these elements and functions not to be found in the system being studied. The danger lies in making the system fit the analogy. In the jargon to be explained below, an analogy is very noisy.

11. Model building

This short section is not intended to give instruction in model design, but to draw attention to the different attitudes to model building which exist. Lowry and Harvey do purport to explain model building, but their examples are very brief and confined basically to mathematical models. The best way to learn model building is to work with a good teacher for a long time.

Harvey describes what he calls the simplest possible model, and what must be part of a 'model-system'. This confusing reference to models within models has been noted above, where the suggestion was made that the mathematical part of a model which needed also a verbal explanatory part should be called a submodel or part model. According to Harvey (*MG*, pp. 552–4), the simplest mathematical submodel needs three types of variables and a set of operating characteristics which link them. For example, to construct a simple model of wages affecting economic demand, there must be the input variables, independent of the submodel, the wages. Then there must be the status 'variables' which are the internal mathematical constants, in this case the wage-earning population; and finally the output variables, the demand for goods, which depends on both the inputs and the status variables, in other words, the demand is a function of the size of the wage-earning population and the size of the wages they earn. The internal functions of the submodel, which are the operating procedures for the computer, can be deterministic, probabilistic, or functional. There may also be a feedback element, in the sense that the size of the demand can in turn have an effect on the size of wages; the more people buy, the larger the turnover, the higher the wages that can be paid, so more people can buy.

Lowry's 'short course in model design' (in Berry and Marble, ch. 5) is also concerned with the mathematical model, but more specifically for the purpose of prediction and planning. Lowry claims that building a model for prediction demands much greater understanding than building a descriptive model. In a descriptive model, he claims, it is sufficient to know that X and Y co-vary in certain ways; but in a predictive model one must know how, why, and which is cause and which effect. In the wages example cause and effect are clear, but in many systems, particularly those studied by urban geographers, cause and effect are not easily identified. Lowry then writes of strategy, fitting and testing the model. First one must choose the type of mathematical operations most suitable to the type of system one is trying to model. This, in itself, can call for as great experience in mathematics as in one of the branches of human geography. Secondly one puts in the numerical data, both the variables and the constants. Again, this is much easier said than done, as the commentaries on research work, complaining of the lack of or difficulty of obtaining the correct data, prove. Finally, one must test the model to see that it does work as planned, before it is used for prediction. Lowry himself admits

113

that it is extremely difficult to find tasks which will test and help prove reliable any mathematical model.

Within these two examples are implied two completely different approaches to model building. One approach is to model the system as it is, or to model it as we believe it is, with our imperfect knowledge and understanding. Thus meteorologists attempt to model the low-pressure system, constructing ever more refined and elaborate models to match the structure and functioning of a depression as closely as possible. The other approach is to model the system as one thinks it ought to be, not as it functions now, in an inefficient way, but as it could function, in a perfect world. For example, many economic geographers are now trying to construct a model for the location of industry which does not represent what industrialists are actually in the habit of doing, but which represents what they ought to do to find the optimum location for a factory. The location of industry in the past has been so much a process of trial and error, blunder, blind good fortune, the operation of social, political, historical, and personal factors that it may well be impossible to model the real system fully. Certainly the model would have to be a complicated affair to represent accurately complex reality.

There are two differences between modelling physical systems and modelling human systems. First, we believe that physical systems operate in a constant, logical way, so that the models can be logical. But human systems are clearly often illogical, in the sense that man is not always affected only by economic factors, and in the sense that emotion may be dominant. So the model, at least, has to be much more complicated and allow for many unpredictable possibilities. Second, there can be no question of modelling how a physical system ought to work, or could work better. Physical systems operate in a way which we may be able to decipher and model, and that is that. In contrast so many of the ways in which man does things could be improved on, and his present methods vary from place to place so much, that perhaps one ideal model is the best thing to aim at for the time being.

Grigg (*MG*, p. 494), writing of the regional analysis as a type of model building, makes the point that the type of model to be built will depend entirely on what the geographer aims to explain. This point may seem so obvious that it is not worth mentioning, yet so much space in *Models in Geography* is devoted to completely different methods of working that the point will be emphasised here. It is urged most strongly that the geographer starts out with a topic or area of interest, something he wants to study, understand and explain and then thinks of a model as one possible tool to help him in this venture. Chorley and Haggett (*MG*), and Bunge (1962) more often than not, recommend the reverse process of obtaining a model and then casting round, looking for something to which to apply it. They are at the same time both convinced of the powers of models, and so well aware of the extreme shortage of good models in geography, that they advocate wholesale borrowing and re-application of models devised for one purpose to, perhaps, a completely different purpose. Models may be helping other disciplines to make rapid progress, geography may have been slow in starting to use models, but a wild scramble to snatch models from other disciplines and apply them to geographical problems just for the sake of having models to apply, may not result in much worthwhile work in geography.

Haggett (Chorley and Haggett, 1970, pp. 112—13) mentions such approaches to model building as simplifying a system down to its essentials, building up a more and more complicated structure, by induction, by *a priori* reasoning, and so on. There could hardly be a standard procedure for such a difficult and personal matter as trying to begin to construct a model of a system not modelled before, but suggested lines of approach such as these help a start to be made on the problem. The processes of original thought are so difficult to understand and explain, and the solutions to problems suggest themselves at such odd times and odd ways, that one would not expect two geographers to have started to construct two models in the same way. What one would expect is for them to have started with a topic of interest and then have attempted to model it in some way.

The first hint of a decline from this is when the geographer starts with a problem, but borrows, rather than constructs, a model. Haggett states that 'some of the most successful have come from borrowing ideas from related fields, especially the field of physics' (Chorley and Haggett, 1970, p. 109). Models, such as the gravity model, have been borrowed from physics, but one is surprised to read that physics is a related field to geography, and the main danger in borrowing models from disciplines rather closer, say geomorphology and economics, is that they will not suit the geographer's purpose. If the geographer discarded borrowed models which did not fit, all would be well, but the evidence suggests that these borrowed models are as dangerous as analogies, in that major features of the borrowed models have no counterpart, in the geographical system, but the geographer reads them in and tries to make them fit.

The inevitable conclusion of a process which starts off in a reasonable way to help add more models to a very small stock is

115

the stage where the geographer has his model, but no topic of interest or problem. Haggett (1965, p. 22) and Bunge (1962), in particular, keep hinting at this state of affairs, and recommend the geographer to go picking over the models of geologists, physicists, chemists, geometricians and others, like thieves, beggars or jackdaws. Such a poverty-stricken approach is described as fruitful, inspiring and stimulating, but one wonders just what kind of stimulating geography is going to be written by a 'geographer' who falls in love with a model in solid state physics and then casts his jaundiced eye over the earth's surface for some unfortunate topic to which to apply it. The geographer, like a specialist in any other discipline, must be aware of the main advances in other disciplines. At times a concept in another discipline will give him an infinitely valuable idea; but one hopes that he is already genuinely concerned with some geographical research, and that he rebuilds the model to suit his purpose. Finally, for the extreme of desperation, there is the shotgun or Pleiades method of trying to generate models which ranks with reading tea-leaves in the bottom of a cup to try to find the way to Lassiter's Reef (Haggett, 1965, pp. 281—6).

12. 'Black box'

The expression 'black box', carrying some suggestion of the mysterious, electronic devices in an aeroplane, could be mistaken to refer both to the newest and most advanced models in geography. Some geographers, especially those interested in simulation and prediction, would certainly like this to be believed, and McCarty and Lindberg almost claim as much. However, Chorley (*MG*, p. 62) shows, more clearly than the others, that the black box type of model is perhaps the oldest, and very often the least accurate, of the types of model used in geography and related disciplines. He writes of three types of model, the synthetic model, the partial model and the black box. The names are not important and are not more jargon to be memorised, but the ideas behind them are revealing. The synthetic model is sufficiently like the real system for one to understand how the system works. The partial model is less like the real system, but helps one to understand enough of the working of the system to be able to predict results from given causes. The black box is a last resort, when the geographer has no idea of how the system works, but he can construct a device, or a mathematical operation which, for given inputs produces the same outputs as the real system.

Chorley claims that the early geomorphologists, Davis, Penck, and Gilbert, used the black box type of model in the early stages of their work. In the early stages they could see, for example, that a change in base level eventually resulted in a new set of landforms, but for a long time the processes in between were not understood; some would claim that they are still not understood. Similarly, farmers throughout the world have known that overcropping and overgrazing result in soil erosion, without necessarily understanding anything of the processes of soil exhaustion and breakdown in between. Thus it is possible to construct a model which represents the causes properly, and represents the effects properly, but in no way accurately represents what goes on in between; this is the black box. The black box can take different forms. Often in the past it has taken the form of an incorrect explanation or hypothesis linking observed cause and observed effect. Tropical swamps were explained as causing malaria by means of the vapours which came up in the night. Input and output were correctly related (where water in the tropics — there malaria) well before the complex process of the breeding of mosquitoes, development of the parasites, and injection into humans via the insect bite was in any way understood.

The more modern form of the black box, especially in human geography, is one in which the geographer does not hypothesise the internal operations between input and output, but devises a computer program which will produce the same results, even if in a different way. The model which simulates the growth of the Negro ghetto in Seattle neither works in the same way as the real process, nor does it offer any explanation of how the real process works. In the days of steam railways, when the driver pulled the lever, the engine moved forward. The modern enthusiast, with his model railway, can also move a miniature lever in a miniature cab, and the engine will move forward; but there is no suggestion either that the electric model works internally in the same way as the real steam engine, nor that the explanation of the working of the electric motor is in any way an explanation of the working of a steam engine.

Haggett (1965, p. 304) describes how corks, fitted with magnets and floating on water, can be made to simulate the hexagonal pattern of the distribution of central places. Uncannily similar as it may be, the rejection of one magnet for another in no sense works in the same way as a town serving its hinterland. Geographers (such as Morrill, and McCarty and Lindberg) who use computer black box simulations in order to be able to predict events from given causes claim that in the end their models *do* work in the same way as the real systems. However, this, at best, can only be to the extent that electrical impulses in a computer are like muddy water in the Mississippi or a Puerto Rican moving into a room in Manhattan. Guelke insists that statistics and mathematical processes will neither explain anything nor throw up new ideas. The black box is very useful for prediction, but its use as a type III model for explanation is very doubtful. Again, the danger lies in trying to explain geomorphological processes or human actions as though they worked in the same way as a digital computer.

13. Noise

One of the most bizarre, and least appropriate, words which Chorley and Haggett have introduced to the jargon of models in geography is 'noise'. The word is so ambiguous that it has to be qualified every time it is used, and it seems to cause confusion out of all proportion to its usefulness. In *Models in Geography* (p. 20) it means random information as distinct from information organised in some regional or topical framework. In *Spatial Analysis* (p. 196) it means random variations, as distinct from the information conveyed in a map, such as a trend surface map. Thus in the first case it means accurate, useful but unorganised information, in the second it means inaccurate, irrelevant or distorted information.

In a separate chapter in *Spatial Analysis* (p. 45), Chorley describes noise in a model as the result of the model building either discarding relevant information or including irrelevant information. Thus one can put in noise by taking out information, as confusing as the larger stop-number on a camera giving a smaller aperture to the lens. Moreover, one can put in noise by putting in the wrong information. Haggett, in *Locational Analysis in Human Geography* (p. 21) mentions only one of the two possibilities described by Chorley, putting the emphasis on the introduction of noise by oversimplification. Board (*MG*, p. 699), in contrast, stresses only the other possibility, that of the introduction of noise by building irrelevant information into the model. According to Board, noise is irrelevant information either put in by the model builder or read in by the model user as the result of his failure to comprehend the model. For him a noisy model can be one which is distorted or one which is difficult to understand and use.

Is it not possible simply to write of random information, random variations, an oversimplified model, one containing irrelevant information, or one difficult to use? Then each meaning would be explicit and a senseless little word would not need to be overworked and qualified until it has lost any connection with what we *hear* and call noise.

7
Problems and dangers of the use of models

Two aspects of the complex nature of models make it necessary to devote a chapter to the problems and dangers of their use, although there is no question of their supreme importance to modern geography. First, one takes it for granted that those who use models will soon see through the error of books on quantitative and other techniques in geography which equate models with a correlation test, with fieldwork or with map-work, and, in fact, devote much more space to the quantitative techniques than to the models themselves (Cole and King, 1968; Toyne and Newby, 1971). If the use of a model were no more involved than a piece of fieldwork, or a series of calculations, tedious but basically simple, there would be few dangers to point out, and not very important consequences of going wrong. Because the model embraces all these techniques, and the much more difficult processes of setting up the hypothesis in the first place and testing the logic of the explanation later, there are many more pitfalls and many more chances for cumulative error.

Second, the model is not only a much more complicated device made up of many more procedures than a statistical test or a field investigation, it is a device of a higher order and of a different type altogether. The model does not just demand first-hand data, and accurate quantification as do the other techniques mentioned, but, unlike them, it necessitates some concept of the interrelationships and function of phenomena, and at least an hypothetical explanation. Thus the successful use of good models in geography offers infinitely more than the mere use of quantitative techniques (which have their function at one stage in the development of the model), it offers the possibility of the understanding of the structure and function of a system on the earth's surface. Not surprisingly then, the greater benefits to be gained are balanced by the much greater risks of making serious mistakes in perception and explanation.

One of the early dangers, stemming from unfamiliarity with models, is the assumption that there is only one model for each

119

type of system. This can be erroneous for a number of reasons. It was noted in the section on normative and ideal models, for example, that there can be more than one ideal, and what one worker is convinced is the only possible, logical ideal arrangement does not impress another. Much more common is the situation where, in the early stages of an investigation, many geographers are not clear what the elements and basic functions of a system are. Different geographers, or one geographer at different times, can and do produce different models of the same system. This, in research, is the basic function of the model, to be set up, tested, modified, tested again, and, most probably, rejected so that a completely new start can be made.

A different danger arises from the reverse situation in which one 'model' seems to fit several different real systems. The most familiar example is the gravity 'model' which fits systems studied in astronomy and in physics as well as in geography. Within geography itself the gravity 'model' is applied to sizes of towns, the sphere of influence, journey to work and now to some aspects of industrial location. The word 'model' has been placed in inverted commas because, of course, the gravity model is only a submodel of function, type II as defined in chapter 2. By itself it has neither the elements, part I, mapped into it, nor an explanation, part III, attached to it. In physics, for example, the elements to be mapped in would be stars and planets, while in geography the elements mapped in are usually towns of different sizes and different distances apart. In physics, the type of explanation given for the observed functioning of the elements is that the gravitational field of one planet acts on the field of another. Now, while the function may appear to be the same in urban geography, both the elements, as stated, are different, *and* a different type of hypothetical explanation is necessary. The attraction of towns is different in type, and operates on people in a different way, psychologically rather than physically, from the operation of gravity.

One danger in such a situation is of trying to apply the same model to every system which appears to have either the same type of structural layout, or the same type of function. A much greater danger is the failure to realise that a complete model is made up of parts I, II and III and either to neglect III or to assume that because II fits two different systems, III must necessarily fit as well. No geographer would be silly enough to think that part III of the physical gravity model fits the human system, that people are dragged to cities by the same irresistible force which makes them fall off ladders, but the need for a different type of explanation between two geographical applications of the gravity model is seldom as clear as this.

In this context, Harvey (1969, pp. 166—7) writes of over-identified, unidentified and identified models. In these terms he is referring to whether or not a model is definitely connected with one, and only one, theoretical explanation. The ideal, of course, is for a model to be identified beyond doubt with the one and only possible explanation. In the terms used in this book, the ideal is for parts I and II of a model to be identified with a part III, and with only one part III. When this ideal has been achieved, we have laws and theories which give the 'proved' description and explanation of a particular phenomenon.

At present, geography abounds with unidentified models. Again the jargon which the model-builders coin is obscure. An unidentified model sounds like an unidentified flying object, some UFO in the sky which we have seen whizzing past at 1 000 miles an hour and do not comprehend. Fortunately the unidentified model is more prosaic, being simply a model with which certain workers are completely familiar, but consisting only of parts I and II. This type of model exists to describe a system which has been observed but for which no explanation at all has been offered. The most familiar example of this is the rank-size rule for towns and cities in a region (Cole and King, 1968, pp. 481—4). These elements are functionally related in such a way that a graph of the rank of each town on one axis plotted against the size of the town on the other axis gives a smooth curve, or a straight line on log-normal paper. This phenomenon is predictable within limits, but as yet no satisfactory explanation has been put forward to account for it.

The overidentified model is that in which there are two or more hypothetical explanations in part III, each of which seems equally plausible; the model is identified with an embarrassing variety of possible explanations. This situation is perhaps as familiar as the unidentified situation, and equally infuriating. In the latter situation, the absence of any attempt to explain the observed phenomenon is mystifying. In the former situation, the fact that there is more than one plausible hypothesis, instead of leading to rigorous tests which will help reject some of them, leads to endless debate and illogical argument as the authors of each hypothesis defend their cherished ideas.

Whatever else the model may be, and whatever the arguments about it, the most important and generally agreed characteristic is that the model is a simplification of reality. Once that point is established, the degree of simplifications leads to as many dangers and problems as any other aspect of the model. The most obvious danger is that of selection and simplification to the point of distortion. This can happen in teaching when a model is already accepted and the system, such as the erosion of landforms, reasonably well understood, when there is the need to simplify for the purposes of introduction. The danger of it happening in research, when the system is not fully investigated and is by no means understood is clearly infinitely greater. Several authorities comment on the amount of abstraction which takes place as models are further simplified. Board (*MG*, p. 697) sees this as an advantage. Writing specifically of maps as models, he states 'as maps become more thematic, more specialised and more quantitative, they become more abstract', and 'the message, although it contains less information, has a much better chance of reaching its destination'. Ambrose (1969, p. 217), in contrast, writing of the model as 'a simplified and perhaps idealised version of some aspect of reality' and as 'abstractions from reality, devised to give us a fuller understanding of reality' realises that they may be distorted until they are misleading.

A closely related problem is whether to construct several simple models of one system, each to represent one aspect, particularly the function of one factor in a system, or to construct one much more complicated model (although still simpler than the real thing) to represent the combined action of all the factors. Each course of action has its merits, and one need not argue in favour of one against the other, the important thing is to point out that the geographer making or using one extreme model or the other must be clear himself, and make it absolutely clear to others, whether he has isolated one factor, or is modelling the operation of many.

Haggett (1965, p. 280) refers to the choice between multiple regression analysis, and the simpler sequential regression analysis which takes one factor at a time. Lowry (in Berry and Marble, 1968, ch. 5) seems to recommend the latter when the aims are understanding of the factors operating in a system, and application of the model to more than one particular real example of the system. McCarty and Lindberg (1966, p. 55), in contrast, seem to take it for granted that the model must be capable of containing every factor which can possibly operate in any conceivable example of the phenomenon in question. For them, a model is 'a hypothesis . . . stated in such a way that the effect of changes in any one of its independent variables can be assessed' (p. 54).

The mention of factor analysis demands a warning, made nearly every time correlation procedures are discussed but, as often as not, disregarded: that the distributions of two phenomena correlate, even perfectly, is in no way proof that the two are connected, and certainly not proof that one causes the other. The correlation can be a matter of coincidence; it can result from the two phenomena being similarly affected by a third factor which has not even been built into the model. Even when a connection is established, the fact of correlation is not an explanation of cause and effect. All one has is a statistical assurance that the combination is significant. However difficult the mathematical operations to reach this point, the much more difficult task remains of hypothesising a chain of cause and effect which could explain such correlation. Geographers should keep this warning constantly in mind.

Returning to the dangers inherent in not defining exactly how many factors the model represents and contains, in addition to the danger of the researcher losing sight of the main issue, and others misunderstanding his purpose, there is the danger that the model will become untestable in a rigorous way. Between the isolation of one factor operating in a system, which is rarely done, and the representation of all the factors operating in a system, which is virtually impossible, lie all the possible permutations of factors believed to be important. In selecting some of these factors to build into the model, the geographer is normally supposed to be selecting those which are important, or those in which he is particularly interested.

Guelke (1971) writes, in this context, of the *ad hoc* theory, by which he means the addition of one or more factors more than the geographer intended, not because he is interested in their operation or because they are important in the view of the world he is trying to establish, but to save his model from being rejected. Guelke is hostile to models, and his criticism needs careful consideration. The example he gives of an *ad hoc* theory added to a model to prevent its rejection during testing is the addition of the factor of rejuvenation to Davis's model of the cycle of erosion.

This is not a convincing example. Guelke implies that the model of the cycle of erosion should have been rejected because the evidence in the field indicates that no cycle is ever completed. In the light of what has been said about models in this book, however, one would argue that the addition of the concept of rejuvenation was a necessary modification of a model in the light of experience. It is not necessary to split hairs and argue whether the model containing rejuvenation is a modified model, or a new model, the point is that it fits more closely to observed reality — and that is what models in geography are for.

What Guelke seems to be criticising is weak scholarship which sets up a model for test and then goes on making allowances for things which should disprove the model, until the model has been changed out of all recognition and no longer has a precise structure or exact quantification. He is arguing for the setting up of models for test in such a way that the results of the test will leave the geographer in no doubt whether his model is applicable to the system or should be rejected and a new one tried. Equally, when the testing makes this clear, the geographer must have the courage to follow the course indicated. In effect, Guelke is arguing against the imprecise nature of so many models being put forward at the moment, models which are so vague that field testing will not reveal whether the model is completely wrong, needs modification in one or two factors, or whether the particular area in which the test takes place has some unique characteristic.

Guelke's other example of the confusion which can arise from the use of *ad hoc* theories is the central place model. This hypothesises that central places are distributed in a triangular pattern, and that towns of a certain size are a certain distance apart. Now Guelke, like many others, would insist that if this model is correct, then there must be no exceptions to this when the model is tested against real towns in Germany, England, the USA or Outer Mongolia. Each time the model is tested against, or compared with, central places in different regions, it does not fit exactly. Towns are not in a perfect triangular pattern; for a certain size they are not always a certain distance apart. The short answer seems to be that the geographer must indicate tolerances for his measurements; instead of saying towns of 5 000 people are 10 km apart he must say, for example, towns between 4 000 and 6 000 are between 7.9 and 13.2 km apart. One can imagine the opponents of quantification beginning to grin.

However, the only danger here is of not sticking to the rules. If the fact that towns of 5 000 are 9.5 km apart rejects the first model, then the fact that towns of 4 000 are only 7.8 km apart must reject the second. This is the crux of Guelke's logical argument, and is a point which so many newcomers to the use of models in geography, unfamiliar with scientific method, cannot see. A much more difficult set of tolerances to establish is the allowable deviation from a triangular pattern. If the towns are at the corners of equilateral triangles, well and good. But if they are off the corners, how far off is slightly off, and how far off is too much off? Guelke's point about *ad hoc* theory would make sense in this connection if, for example, the geographer building a model showing towns at the corners of equilateral triangles, in order to protect his model from rejection at all costs, calmly added a few right-angled triangles to make the model fit a few real cases.

124

In the end of course, the central place geographer has an escape clause which must be the envy of every lawyer in the land. He claims that the central place model represents only what would be the system of central places on an isotropic surface. Now, as Guelke (1971, p. 47) says, 'there are no isolated city states or perfectly flat plains extending in all directions', so, in effect, the central place geographer can attribute all variations of distance between towns, and all variations in the triangular pattern to the effects of the shape of the country and the variations in the relief. An easy way out, but it misses one of the main purposes of models in geography. In geography, as distinct from economics, regional analysis, or any other discipline, our understanding, in the end, must be of the real earth's surface. Therefore it is contended that when the geographer makes use of the central place model he must use it to help understand real landscapes. Therefore the central place model in the future must be developed so that it can *measure* the exact effect of the shape of countries, of relief, rivers, variations in soil, the locations of break of bulk and resource-based towns on the basic central place distribution. This brings us back to the original point of models which study either the operation of one factor at a time on the distribution of such things as central places, or the operation of many factors in combination. Even slight consideration of the problems of measuring the influences of all the factors on the distribution of central places will suggest that the early models should attempt to evaluate one factor at a time.

Keeble (*MG*, ch. 8) suggests three criteria for a model. He says that the model should single out crucial variables, define its concepts precisely enough to permit comparison with the real world, and possess an analytical rather than a descriptive framework. One would qualify this by insisting, first, that it singles out a stated number of crucial variables, they must not be like the damping rods in an atomic pile which can be added or subtracted to stop the mass going critical. Second, rather more firmly, one would argue not just for precise comparison with the real world but such rigorous testing that the test either clearly rejects the model, or gives a sure, acceptable indication that it represents the truth. 'Without built-in testability a theory or model is of no more explanatory worth than an astrologer's calendar, even when it is compiled on an IBM computer' (Guelke, 1971, p. 48).

In modifying King's classification of models earlier, reference was made to the fact that each geographer, carrying out research into a particular branch of geography, has a rather specialised concept of models, influenced by his own work and ideas. Particularly for the student, this can lead to confusion when familiarity with models in the early stages is gained only in one branch, and a one-sided view develops. For example, while King refers to models very much in terms applicable to geomorphology, Ambrose (1969, p. 217) seems to take it for granted that all models are predictive.

This minor danger can soon be overcome, but there is a much more serious aspect of individualism. Geographers not only have different concepts of what constitutes a model, but they also use them in different ways, for research, for prediction, for teaching and so on. Moreover, even when Haggett's recommendations are followed with caution, there will be a fair amount of borrowing from other disciplines, and it will be even more important to realise that specialists in other disciplines have different concepts again of the model, and that they use them for many different purposes. At best, inexperience can lead to the danger of the geographer misunderstanding a model, or misapplying it, in the extreme case, of course, pursuing a course of action which has no connection with geography at all. Above all, Guelke seems to fear that hypothetical models, still a long way from being tested, accepted and identified with theory, are put to practical uses by what he calls social technologists. This parallels the common practice of second, third, fourth, etc., geographers eagerly accepting the hypothesis of the first as proved fact.

The same idea is inherent in the use of black box models, with one vital difference. Many of those who use black box procedures for prediction and planning are careful to point out that they do not understand why a system works, only that it *does* work in this way and can be represented and predicted. It is argued that the work of the geographer and the planner are essentially different in this respect. The planner does not need to understand a system if he can reproduce it in a model and predict changes and predict the effect of changes. The geographer must aim to understand. Similarly, Mesarovic, quoted by Chisholm (1967), complains that the engineer (social technologist in Guelke's terminology) is looking for general concepts which in the short run will help them put things into practice, not for principles which are eternally true. Both aims are valid, one merely hopes that geographers will not confuse the two.

Harvey (1969, p. 163) writes of geographers abusing the model concept. He argues that in geography, where theory is weakly developed, the use of *a priori* models is inevitable and he points out that the inexplicit use of *a priori* models, or hypotheses, is particularly dangerous because 'a curious transformation may take place from an *a priori* model which has no

126

empirical justification, through the sudden acceptance of that model as *being* the theory (without any empirical evidence), to the ultimate canonisation of the model as the quintessence of reality itself'. Thus Davis's hypothesis about the cycle of erosion was accepted as truth by D. W. Johnson, and Ratzel's hypothesis about determinism was accepted as truth by E. C. Semple.

Many people believe that computer models are the most accurate and reliable, and prediction the most useful application of models. Guelke (1971) reminds geographers that, for prediction in a scientific manner, they must know (*a*) the determining conditions of the events to be predicted and (*b*) the laws governing the system about which predictions are to be made. He believes that accurate prediction is unlikely in human geography because the determining conditions are just not known, and because he believes the 'laws' can never be more than weak generalisations. Even in physical geography, where the systems operate more regularly, those who study such things cannot even predict the occurrence of earthquakes because, while some laws governing their functioning have been formulated, they have not yet discovered the determining conditions.

In the same vein, where more of the laws and some of the determining conditions are known, Manley (1971) felt it necessary to apologise for the weather forecasters making so many inaccurate forecasts in 1971:

> We cannot rely too much on the notion that the weather patterns, represented by distribution of anomalies of pressure and temperature at a given season in previous years will necessarily reproduce or repeat themselves. There are too many variables, and it begins to look as if the sea surface temperature is one. Moreover, we have only about eighty years of 'monthly patterns' to go on; even finding a close analogue for any particular month is not easy.

It was almost impossible to resist the temptation to put the whole quotation in italics and to use it to drive home other points made in this book. In this matter of prediction there are three elements, the determining conditions, the governing laws, and the event to be predicted. Like the old problems of triangles in geometry, one must know two sides or two angles *exactly*, before one can predict the third.

In the concluding section of this brief survey of some of the problems and dangers of model use, three ideas will be considered together. These are that the reality we observe is not the only possible system which could have occurred, that models are not universally applicable, and that the model is not reality.

127

Regional geographers sometimes fell into the error of assuming that the landscape and the economic activities which they observed were the only possible arrangements which could have developed in that place. General geographers, using models, may be more scientific, but too often it is clear that they set about modelling and explaining a system as if that were the only possible kind of system and the only way it could operate. Chisholm (1970, p. 105) is the authority most explicitly aware of this danger: 'In explaining the distribution of geographic patterns, one should consciously think of the field of probabilities at each stage, in an attempt to assess the extent to which the actual pattern deviates from a truly random one.' Any model must make provision for considering alternative patterns, not assuming that what did develop on the earth's surface was the only thing which could have developed. Thus Chisholm, referring to the location of industry, emphasises that one should, for example, consider not only why steel located in the Clyde valley rather than other valleys, but why steel developed there rather than any other industry.

'Lösch, in particular, develops his theories regarding the nature of the ideal distributions and then seeks evidence that in fact reality does conform' (Chisholm, 1966, p. 15). Any use of an ideal model, must inevitably condition the geographer to thinking that only one system, his ideal, is possible. If an authority such as Lösch can fall into this error, as Chisholm claims, then the novice must be very careful indeed. One of the major obstacles to geographers ever giving a true explanation of the systems on the earth's surface is the train of thought which starts by assuming that what we see was inevitable from 4004 BC, and ends by thinking that the description of the sequence of events between then and now is an explanation, even a geographical explanation.

Of the many reasons given for geography's change to a model-based paradigm, the most important single reason is that this will allow laws to be formulated. When regional geography was predominant, the interest was in the special and the particular, it was impossible to formulate laws; at best, geographers could only make generalisations about the particular regions which they had investigated. With general geography now predominant the interest is in the general and the universal, and the belief is that models will help in the formulation of laws which are necessarily true in those areas which geographers have not investigated, as well as in those that they have. The advocates of general geography, using models, denigrate the old-fashioned ideographic geography, giving it the derogatory label of particularism, and imply that the models and laws to be established in the branches of general geography will be universally applicable to every part of the world. This may prove to be the case in physical geography, but the workers who are getting on with the job of research into human geography are not so sure, as will be shown below. However, Harvey (1969) stresses the same basic idea that the laws which geographers hope to establish with the aid of models should be universal in space and time; he then modifies this to methodological universality, that is, all over the world, rather than throughout the universe (p. 101).

In contrast to these sweeping claims for the universal applicability of models, one finds words of caution from workers on particular topics. Lowry (in Berry and Marble, 1968, p. 61), envisaging the time when a simulation model of the economy of a city has been perfected, insists that the model will fit only the city for which it has been built, and that a new simulation model must be built for each other city to be studied. McCarty and Lindberg (1966, pp. 9, 46) sustain a similar kind of argument over many pages. In explaining why there are branches such as economic and social geography, they make it clear that inevitably one model will not apply to the whole world because of the many possible permutations of factors, even in one type of system in economic geography. Harvey (1969, pp. 90, 96), having insisted, correctly, that laws must be universal, points out the practical difficulties, particularly in the social sciences and human geography, of relating the model and the theory to particular phenomena.

Much later in his book Harvey returns to this problem, when he emphasises that conclusions from a test on a given population (population in the mathematical sense of the word) refer *only* to that population (p. 282); thus even when geographers are working toward general laws, they will have to test either the majority of particular world populations or a sample of the worldwide population. He shows how the test of a model in one area can be misinterpreted in five different ways, and implies the wider the extension of the conclusions to other parts of the world, the greater the accumulated errors.

Haggett (Chorley and Haggett, 1970, pp. 361–8) admits a problem of the same order, in a different context. Writing of the G-Scale he implies that different types of model may be necessary for work at different regional scales; for example, a model developed for one county on the High Plains might apply to other counties on the same G-Scale but not to the whole of the Plains which is a region on a different G-Scale: 'No one model should be expected to accommodate many aspects of reality, at different levels of information content, sophistication and time.' One is tempted to add 'and place' but Haggett is not quite contradicting himself here. The problem referred to is that of the amount of simplification in the model, and Chorley and Haggett seem to disagree with McCarty and Lindberg on this.

Not only McCarty and Lindberg appear to disagree; several contributors to *Models in Geography* disagree with the belief of their editors, Chorley and Haggett, that models are universally applicable and will enable geography to become scientific and nomothetic. Brief quotations are given below (all from *Models in Geography*), to indicate the variety and extent of the feelings that models in geography are not universally applicable but the reader is urged to consult the complete chapters.

Barry (ch. 4) refers to 'synoptic models in mid-latitudes and the quite different ones developed for the tropics' (p. 99); later he explains the necessity for this because tropical models are 'frequently biased by ideas developed in middle latitudes' (p. 118). Not only are temperate models inapplicable to the tropics, but it seems that different models may be needed for different parts of the tropics when Barry writes, 'attempts to apply the existing models of wave disturbances in African forecasting have generally failed and this has led to quite different ideas from those developed in other sectors of the tropics' (p. 126).

More (ch. 5), writing about hydrological models, states, 'Penman's formula is limited in its applicability to British and Western European climatic conditions' (p. 158). She believes that different models of potential transpiration will be needed for both different latitudes and for the edges and centres of continents, for example for the west, centre and east of the USA.

While Wrigley (ch. 6, p. 191) argues at first for the study of the general rather than the special in demography, he later implies that at least two models are necessary, one for 'Western' societies and one for the societies of the developing countries (p. 209). The conventional use of the word 'Western' should not obscure the fact that this means one model for midlatitudes, and a different model for most of the tropics and subtropics. Moreover, like Barry, Pahl (ch. 7) hints at subdivision within the tropics when he quotes Gluckman and Devons as writing: 'Those who engage in this activity should not arrogantly assume . . . that a model which appears to explain development in one country will have universal application and validity' (p. 240), i.e., in other developing countries.

Keeble (ch. 8) contradicts this by asserting that under-developed countries are 'dominated by common features and problems' (p. 244); he then spends four sides attacking the ideographic approach and one page effectively recommending it (pp. 243—6). To add to the confusion Keeble goes on: 'The nature of problems of economic development in a large country are often significantly different from those of a small one' (p. 254), and advocates 'the development of more limited models . . . referring to particular groups of countries' (p. 655).

Garner (ch. 9), reviewing models in urban and settlement geography, and advocating the identification of general systems, warns that, 'the difficulties of generalising over large areas such as the entire USA is (*sic*) also clearly revealed by the marked regional differences in spacing (of central places)' (p. 314); later he admits 'the model gave a better fit when regional differences were considered' (p. 315). 'The area in which Weaver developed his (model), the Mid-West of the United States, was particularly suited to this', writes Henshall (ch. 11, p. 440). By the very nature of models, if they are to be universal, then one area cannot be more particularly suited to the model than another. Henshall's implication that Weaver's model is less suitable to other areas thereby implies that it is not universal. Harvey (ch. 14), writes in the same vein, 'in many cases special factors, such as . . . make it seem almost impossible to set up theoretical models of settlement development' (p. 566).

Harvey, being even more specific, writes 'we cannot make the same assumptions as regards traffic flow (as in physical processes) where technology, cost, motivation, and behaviour may change radically from *time to time*, and place to place' (p. 587; my italics). Harvey, here and in his book, shows as much contrast between the researcher faced with practical problems and the sanguine theorist, as do two completely different people in the one book, say Barry (ch. 4) and Haggett (ch. 1). Finally, Morgan (ch. 16) tells us 'it is a wise precaution constantly to test one's model against the real world, because each situation contains elements of uniqueness' (p. 741).

Campbell and Wood (in Cooke and Johnson, 1969), reviewing the development of theory in human geography, warn geographers not to apply the same model to different times and to different cultures, because different factors apply. Moreover, in other applications, they realise that 'several interrelated models will have to be used' (p. 86). Chorley (Chorley and Haggett, 1970, p. 24), thinking that he has defeated the claim that Davis's cycle of erosion model is universal, thinking, in fact, that he has rejected the model once and for all, actually turns the argument against the universality of models in general, and states, 'no single model can form a universally appropriate approximation to a segment of reality'.

In these statements, one notices, first of all, that people using models for research now realise that different models will be needed for different latitudes, particularly for midlatitudes and the tropics. One model for the developed and one for the under-developed countries is merely a variation on this theme. Further, some of them imply a model for each continent in the tropics, and one for large countries, one for small. Then, such phrases as: 'particular groups of countries', 'marked regional differences', 'particularly suited', 'special factors', 'change radically . . . from place to place', and 'each situation contains elements of uniqueness', must give one pause for thought because these phrases are used by general geographers employing models. On reading these last few paragraphs the regional geographer may well fill his pen and think of coming out of retirement. However much this may have raised his morale, the main purpose here is to warn those who do have to work with models in any way that the evidence shows that geographers are still a very long way from building models which are universally applicable.

Guelke (1971) attacks models on similar grounds, although less specifically. He argues that there are no laws of human behaviour in a geographical sense because there are different uses of the earth's surface in different cultures, and because man learns from the past. He echoes some of the authorities above who imply a different model for each culture area. He weakens his own argument by hedging his bets. He goes on to say that even if laws are found, first, they will be so superficial and general as to be worthless; second, even if significant laws are found (he retreats stage by stage too easily) they will not necessarily be geographical laws or relevant to geography. Some parts of Guelke's paper are very good, but he falls into the error here of arguing that general laws can not be relevant to geography because he has defined geography as concerned with regional differences and interested in special cases. He says geography can not be scientific by already having defined it as non-scientific. With this reservation, however, Guelke and the others raise enough points for one to treat the claim of universality with extreme caution, at the present time.

On reading the simple statement on the page that the model must not be mistaken for reality, those not familiar with models may wonder how on earth anyone could make such a mistake. Unfortunately, experience shows that this mistake is made all too often, and those most experienced in the use of models, and in investigating their nature are most concerned to give the warning. In extreme cases one meets the ridiculous situations of, say, where reality and model do not fit, of the model being defended and the particular area of reality being ignored or rejected because it does not fit in with a model-induced idea of what reality ought to be like.

The main practical danger lies in the fact that the model represents only the elements and the functioning of the system. The model may have an explanation attached to it as part III, but a test of the model is only a test of parts I and II, it is only a test of the results and effects of the functioning of the system. The test of the model, however, successful, is not a test of the causes contained in the explanatory part III. So even a well tested and accepted model is a good explanation of reality *only* if it was a good model for the theory in the first place. Harvey (1969) needs so much space to begin to make this clear, that in an introduction one can only mention the point, and insist that the serious student master what Harvey has to say (especially p. 164ff). For example, he shows that because a stochastic model simulates reality this does not prove that the world is governed by the laws of chance. Similarly, because a cause and effect model can successfully represent other aspects of reality, neither does this prove that the world is ruled by direct cause and effect determinism.

> The empirical success of a theory relies entirely upon the success of the *text* in linking the abstract symbols of the theory to real world events. As Kemery . . . has pointed out, 'establishing a connection between these two worlds is one of the most difficult tasks a scientist must face' (p. 90).

Chorley (in Berry and Marble, 1968, p. 44) has drawn a 'model for models' reproduced in modified forms in several works as well as the present volume. Above all, the point to be stressed about this is that the whole algorithm starts with a segment of the real world, and ends with conclusions about the real world. Chorley is in no doubt, and makes it clear diagrammatically, that the geographer uses a model to study 'a segment of the real world', and not to produce beautiful models.

This list of some problems and dangers of the use of models may be alarming, but it is by no means complete. Just as one hopes that some chapters in this book will help the student to understand models and begin to be able to use them properly, one hopes that this chapter will prevent some of that blithe, overconfident use which stems from lack of appreciation of the deeper problems.

8
Models and the aims of geography

An examination of the use of models in geography is implicitly an examination of major aspects of the methodology and aims of the discipline. Experience has shown that, among students, models become confused with other techniques and methods, while geographers take it for granted that only one aim of geography is under discussion. Therefore it seems necessary for a concluding chapter to make some comment on the relationship between models and quantitative techniques, and between models and the aims of geography, in order to get a proper perspective on models as a whole.

Discussion is often confused by the two common assumptions that models and quantitative techniques are one and the same thing, and that quantitative techniques have made geography more accurate and truthful. A nice distinction must be made in the matter of this second assumption. Quantification is not simply a matter of putting figures into geography, of saying 89 per cent of the farmers in Cheshire are dairy farmers instead of asserting that the vast majority of farmers in Cheshire are dairy farmers. Quantitative techniques are not only designed for greater factual, numerical accuracy. Above all, they serve to test deductions from the hypotheses put forward. The revolution is not in the figures and formulae spattered over every page, but in the fact that hypotheses are rigorously tested before they can be accepted.

The quantitative revolution in geography has three aspects of increasing complexity. First is the simple introduction of accurate numbers into descriptive statements, for example saying that wheat brings in 57 per cent of income, instead of that it is the main crop. Second is the use of statistical techniques to analyse complex sets of data and the effect of multiplicity of factors on a given phenomenon. Third, and most important, is the use of accurate, complete, up-to-date, and relevant numerical data to test deductions from hypotheses, and thereby to test the models. The introduction of numbers at stage one is useless if the explanation remains dubious because it is still untested; thus only stages two and three result in genuine progress.

Another misapprehension, closely connected with this misunderstanding of the role of quantitative techniques, is that quantitative techniques have removed from geography both the qualitative element and the need for imagination. Nothing could be further from the truth. The qualitative stage of the work has merely been transferred, while the need for original thought, for imagination, is greater than ever. There may no longer be room in geography for qualitative statements which have not been verified and are not supported by statistics, but there is a growing need for qualitative thought and work in model building, and in setting up the hypotheses in the first place.

The use of mathematical techniques to test models has eased some problems while aggravating others. Geographers are now less open to the criticism of being imprecise in their statistics, but more and more are required to give a valid, acceptable, functional explanation of the phenomena they observe and the data they collect. As Guelke (1971) insists, statistics will show patterns, but will not explain them; the geographer has to search for the explanatory process, which is qualitative work, demanding imagination; it is too easy to describe complex patterns in mathematical terms without any understanding of the processes involved.

In fact, for Lowry (in Berry and Marble, 1968), the model is divided into the specific, descriptive, quantitative parts (I and II) and the general, explanatory, qualitative part (III). Moreover, above and beyond the specific points made, Lowry's article, describing the strategy of model design, fitting the model to the data, and testing the mathematical model, shows just how artistic the work is. Any non-mathematician who imagines that computers have solved all the geographer's problems must read this article to see how much imagination, guesswork, compromise and other qualitative attributes go into seemingly the most precise, quantitative work. Pahl (*MG*, ch. 7) goes further than Lowry in not just distinguishing between the quantitative and qualitative aspects of the work, but in arguing for conceptual, non-materialist value models in helping to explain such things as the distribution of different types of people in towns. Grigg (*MG*, p. 499), admits that some aspects of the old qualitative work are still necessary:

If we regard regionalisation as a classifying process, then we immediately commit ourselves to harnessing the use of computers and the increasing precision in the establishing of similar classes. . . . We end up with elegantly designed systems

135

of regions which may have, however, little to offer in the way of explanation. It may well be that the imperfect, deductive and genetic regional systems, however they contravene the rules of classification, may be the more stimulating vehicles of inquiry.

One begins to understand why some models appear to contain no element of explanation at all.

Hamilton (*MG*, p. 368), perhaps more hopeful that quantitative techniques will help in explanation, is still aware that they can not account for everything when he writes: 'While, therefore, the beauty of simple, economic interpretation is the attraction of a model's vital statistics, the reality of her heart reminds one of the dangers of ignoring aspects which can not be expressed in figures!' Chisholm (1966, pp. 15, 16), too, is concerned with this aspect of the quantitative revolution, that some vital factors can not, at present, be quantified. Several research workers admit this, and explain that they have substituted other data, or have inserted purely arbitrary figures. This, of course, just will not do. To substitute factors, to assign arbitrary values, or to ignore factors altogether because they can neither be quantified nor represented by something else is infinitely worse than not pretending to quantify in the first place. The two grave dangers here are, first, of making something inexact seem to be precise, and second, of ignoring some aspects of geography entirely.

Many of the books which aim to introduce students for the first time to quantitative techniques* tend to treat these techniques in isolation from models. Mainly for reasons of space, and mathematical clarity, the quantitative techniques are not emphasised, as they should be, as part of the work of testing an hypothesis. In fact very often it is overlooked that there is an hypothesis in the first place, and so the impression in the student's mind is that one starts, for some obscure reason, with a mass of data which has to be sorted out. Perhaps even worse, some books, or some of their sections, give the impression that quantitative techniques and models are separate, but the same kind of thing, of the same order of importance. For example, some books include the rank size model, the gravity model and Huff's probability model, and appear to give them similar treatment to sampling or correlation; yet these three models named, and others are, in fact, models of systems and not techniques for collecting and comparing data.

---

* Berry and Marble (1968), Chorley and Haggett (1969), Cole and King (1968), Gregory (1973), Haggett (1965), Theakstone and Harrison (1970), Tidswell and Barker (1971), Toyne and Newby (1971), Yeates (1968).

The publication of a dozen or more such books in the last decade has helped to foster the idea, widespread in some quarters, that the techniques have been introduced into geography in order to deal with an increasing deluge of numerical data which in some mysterious way geographers have decided to analyse. One must question this on two grounds. First, there is not a sudden wealth of numerical data; and second, if the techniques were introduced only for the purpose of geographers suddenly coping with an unexpected flood of data geographers would not know how and when to apply them, nor have any valid reason for doing so.

Henshall (*MG*, p. 452), describing Beguin's fascinating model of tropical agriculture, has to dismiss it more quickly than one would like, when she writes 'its main disadvantage is the difficulty of obtaining the empirical data on which it is based'. Haggett (*MG*, p. 655), writing of the analysis of traffic flows on railways, admits that this 'demands data that often are simply not available'. Keeble (*MG*, p. 270) complains of 'the enormous difficulty of obtaining the detailed regional data needed for input—output analysis'. As recently as 1969 Grigg wrote:

First, those countries that have adequate means of collecting data on agriculture do not always collect information on those things that would be most useful . . ., nor do they always publish the data for suitable administrative units. Second, outside North America and Europe reliable data are hard to come by. Thus, there is little likelihood in the immediate future of basing a system of agricultural regions of the world upon comparable and comprehensive statistical information.

Another example is to be found in Estall (1972, pp. 42, 62—3).

In the light of such statements, anyone criticised for not using quantitative techniques when so much data is supposed to be available, must demand that the critic supplies these elusive data.

One would argue that there is no such thing as a data revolution in geography, certainly not in the sense of a sudden large flow of numerical data of exactly the kind, and in exactly the form that geographers want in order to continue to pursue long established aims more effectively. There is no data revolution in the sense of the *required* data suddenly being made available; and there is no data revolution in the sense of geographers suddenly starting to analyse types of data which they have never required before, simply because now, like Mount Everest, it happens to be there. If this were so, geographers would stand accused of being deflected from their aims by irrelevant details, instead of collecting the data they really need for their purposes.

Paradoxically, the quantitative techniques now employed can make geography less precise than in the past; or, more truthfully, they can serve to show exactly how imprecise so many statements are. A brief outline of the functions of the main techniques, listed below, should help to make this clear. It is rarely possible to obtain data for the whole population, or even to handle such data if it were available, so the first process usually is to take a sample. Then the sample is generalised to such statistics as mean, mode or median, and to standard deviation or variance. Even when the whole population is used the values are generalised to the parameters of the mean and the standard deviation from the mean of the whole population. One of the commonest techniques to be applied to sample statistics or parameters is some form of correlation, and the coefficients obtained will vary between minus one and plus one. With correlation coefficients it is necessary to look up tables to assess the statistical significance of the results obtained. Some results can be accepted with 99 per cent confidence, others with only 95 per cent confidence — that is, five times out of every hundred the results obtained could have happened completely by chance.

Thus when one realises that so many quantitative techniques involve sampling, generalisation into mean and standard deviation and correlation coefficients which may, on occasion, be accepted at +0.7 with only 95 per cent confidence, then one must also realise that the final statement is also a generalisation. Perhaps what these quantitative techniques really do is enable us to say with great precision, just how inexact our generalisations are. Thus nowadays, when a precise figure is given as an average or mean, the geographer is also required to give at least the standard deviation, so that it is clear how many cases will vary by a predictable amount from the mean.

14. Quantitative techniques commonly listed in introductory textbooks

A. General

TO SELECT DATA
Random sampling
Stratified sampling
Systematic sampling
Stratified, systematic, unaligned sampling
Nested or hierarchical sampling

Quadrat sampling
Circular sampling

TO GENERALISE DATA
Calculation of mean, mode, median
Calculation of standard deviation and variance
Mean deviation
Coefficient of variance
Running mean
Index numbers
Standard scores

CORRELATION
Spearman's rank correlation
Pearson's product moment correlation
Co-variance
Regression

TEST FOR SIGNIFICANCE
Chi-square test
Kolmogorov Smirnov test
Mann Whitney U test
Snedecor's Variance Ratio test
Student's *t* test

CLASSIFICATION
Chi-squared for classification
Distance analysis in *n*-dimensional space
Analysis of variance

B. **SPECIFIC**

POINT ANALYSIS
Mean centre of dot distributions
Standard distance
Nearest neighbour
Distance index for farms
Population potential

NETWORK ANALYSIS
Alpha index
Beta index
Pi index
Smeed's index
Diameter of networks
Density of networks
Shortest path matrix
Route shape index

FOR SOCIAL AND ECONOMIC DATA
Coefficient of association (or linkage, or similarity)
Index of specialisation
Location quotient
Lorenz curve

FOR SETTLEMENT DATA
Rank size rule
Gravity model
Boundary girdles and Thiessen polygons
Break point formula
Reilly's law
Huff's probability model

The grouping under headings in this list is not meant to be a definitive classification, but simply to indicate the main function of each of the techniques. Moreover, as the list is compiled from the contents of the most readily available books which aim to introduce students to the techniques, there is no claim that it is complete; moreover, because these techniques are the ones first encountered by the student, many of them are only the necessary preliminaries for preparing the data before more complex calculations are carried out. These are by no means the only techniques used in hypothesis testing, nor even the most important. Increasingly, the techniques which are used to test models necessitate a computer and some lengthy programming, not only because they are complex, but also because of the enormous number of repetitive calculations which must be carried out. Berry and Marble (1968), and Yeates (1968) provide good introductions to the relevance of computers to the work.

The first part of the list is headed 'General' because the techniques have several different functions and can be applied to data from both human and physical geography. Under B, headed 'Specific' the techniques have more specialised functions. However, the point analysis and network analysis techniques can have general application once the phenomena such as population and settlements have been reduced to dots, or the roads, railways and air routes have been reduced to lines.

The last ten items in the list are the most specific of all, and here one can see the contrast between, say, random sampling which can be used for any data, and the break-point formula which applies only to the urban field boundary between towns. Thus further confusion may be caused in these introductory works by mixing specific and general techniques. Moreover, there is extra distortion and confusion in that the last ten items in the list, rather than being complete techniques in their own right, are parts of greater wholes, sub-parts of larger models, presented to the student to a greater or lesser extent out of context, divorced from the rest of the models to which they belong. The sample of books from which the list was compiled may well have been biased, partly by what happens to be readily available, and partly by the author's interest, but another feature is the preponderance of techniques specifically related to human geography over any specifically related to physical geography. When one stands back to review the field this is not serious, but when a student is trying to master these techniques and models for the first time, his concern with individual trees may prevent him from seeing the wood as a whole, and divert his attention to those techniques which just happen to be presented to him first, rather than to those which he needs for a particular piece of work. While granting that the experienced geographer is less likely to be diverted in this way, one would just mention at the present state of development, when models and quantitative techniques are *not* equally well developed in every branch of geography, there are instances when one gets the impression that some geographers have adapted their aims to fit the techniques which have come to hand, rather than searching for, or devising, techniques appropriate to what they want to do.

In this connection it seems pertinent to point out how very few of the quantitative techniques presented to the student geographer are in fact techniques of *spatial* analysis. This is even more disappointing at a time when the spatial attributes of geography, the study of space relationships and relative location, are being stressed as the essential quality which makes geography a separate and distinct discipline. Many of the techniques were developed for work in other disciplines, so that even such useful and essential techniques as sampling and correlation are more often carried out on lists of numerical data rather than on data presented spatially in maps and air photographs. Sampling from maps is done, of course, as is correlation between two sets of distributions, but the techniques which the geographer has to use are very cumbersome modifications of techniques developed for other purposes. The specific techniques, such as standard distance, nearest neighbour analysis, gravity model and break point formula are, of course, directly applicable to map analysis, but even the list under heading B is misleading in this respect, because such things as location quotients work only with tabulated data, and the eight techniques for network analysis are all variations on the theme of connectivity. In the development of models and of quantitative techniques, however, there is very rapid evolution if not revolution, and the reader is urged to consult the most recent journals, where research workers are explaining their new models and their new techniques, as well as their findings about some part of the earth's surface.

The quantitative techniques presented in isolation to the student are unlikely to make complete sense until it is realised that these are normally used in the process of developing a model or testing an hypothesis. The development of such techniques is not a reaction to a flood of data which just needs 'processing'; on the contrary, it is a necessary adjunct to the use of models, and usually results in the research workers having to go out and collect exactly the kind of data they need.

The chain of events is not: flood of data — illogical desire to reduce it to statistics and quotients — adoption of techniques — resultant revolution in method. The chain of events is: revolution in the approach to geography — adoption of a new paradigm — need to test models — adoption of necessary techniques — often vain search for the relevant data. It is precisely because quantitative techniques have been introduced to geography to test the hypothetical models that geographers do know exactly how and when to apply them, and what to do with the results. The research geographer is thus in a completely different position from the student and teacher who have mastered the techniques but have neither the necessary wealth of data, nor the full conceptual model, for the techniques to be relevant to their work.

Quantitative techniques result, therefore, from a change in geography, they have not caused it. Neither have they removed the qualitative element from geography. Cooke and Johnson (1969, pp. 84—7) go as far as to say that the two are not separate; quantitative techniques can only test an hypothesis (see also Theakstone and Harrison, 1970, pp. 97—100). The fact that the hypothesis must exist in the first place argues both that there is a qualitative element in geography *and* that quantitative techniques fit inside a much larger structure. This much larger structure is the complete model. On the relationship between models and quantitative techniques, one must conclude, therefore, first that the use of the techniques results from, and depends on, what the geographer is trying to do, as embodied in his model; second, as demonstrated in Fig. 8.1, that the application of the technique is only one stage in a whole programme of work, again implied by the nature of the model. The model is vastly more complex, and infinitely more important, than the statistical techniques which it embraces.

Any consideration of the relationship between models and the

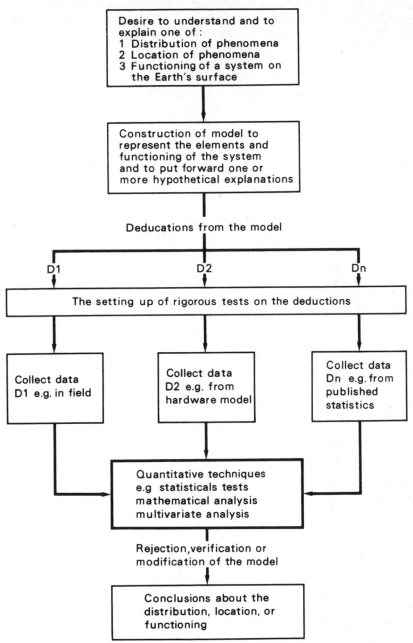

Fig. 8.1 The relationship of quantitative techniques to models in geography. Quantitative techniques do not make sense unless structurally part of a model

aims of geography must take note of the extreme divergence of those aims at the moment. Harvey (1969, pp. 51—2), simplifying to the point where he must have annoyed some historians and natural scientists, points out that while science aims to formulate general laws by means of the study of particular cases, history aims to study particular events and movements and explain them by reference to general laws of human behaviour. The two elements of general laws and particular cases are present in both types of work, but one discipline is interested in the laws, the other in the particular cases themselves. This may be doing an injustice to some historians, but Toynbee in his *Study of History* seems to agree with this, and the point to be made here is that two such different aims are embodied in two separate and distinct disciplines.

In different places, in the work of different geographers, possibly with the emphasis at different times, work which is called geography embodies *both* these aims. Not only regional geography, but those branches concerned with the evolution of the landscape, and with man—land interrelationships now called human ecology, all study particular areas, particular cases. In contrast, most branches of general geography, and the dominant spatial distribution school all aim to formulate general laws. Not surprisingly, it becomes increasingly difficult to conceive of all this work, with such divergent aims, as belonging to the same discipline. Hartshorne's (1939) view was that general geography could be preliminary to regional geography, and when sufficient laws had been formulated, geographers could study particular parts of the earth's surface with reference to these laws. Haggett (1965), however, insists that regions are just the testing grounds for general laws, and that the formulation of laws, not the study of particular parts of the earth's surface, is the aim of geography.

Thus the aims of different parts of geography can be as different as the aims of history and science, and the application of models must be viewed in this light. Harvey (1969, pp. 114—16) describes five themes or aims of geography which will be mentioned here without any implication that these are the only aims, or that they are all equally important at the moment. They are the study of areal differentiation, the evolution of the landscape, the relationship of man and land, spatial distribution, and the geometric aspects of the earth's surface. With such diverse aims, it is not just a question of which can be pursued by means of models and which cannot, but also of which are being pursued actively at the moment. In one case, at least, the use of models could give considerable progress in a field where geographers show little interest today.

Thus at first glance it could appear that the study of areal differentiation is both out of fashion and not suitable for the application of models. Regional geography is still being written, and it is still being studied in most schools and colleges, often under the name of area studies because region is such an embarrassing word. However, the reason that models have not been applied to the study of formal regions is because most geographers have never been clear just what it is they want to understand about formal regions (see Minshull, 1967). This point cannot be too strongly emphasised, and is one of the most important conclusions about the use of models in geography: models cannot be applied successfully until the geographer is absolutely clear exactly what it is he wants to understand. This seems to be difficult enough for some research workers, but it is made much more difficult for all those teachers whose syllabuses are made up of bits and pieces accumulated from all five of the themes mentioned above.

Models could be applied with great effect to the study of functional regions. This is clear enough because the functional region, as defined, is just the kind of spatial system that is being analysed most successfully in general geography. Three things make the application unlikely in the near future. First, the declining interest in regional geography, or the fear of admitting interest in the present climate of opinion generated by the mandarins. Second, the fact that functional regions are the most complicated systems which geographers have ever claimed to study, and their analysis may have to await greater knowledge and better techniques. Third, again, regional geographers have rarely made it clear exactly what they want to know about the functioning of regions.

The study of areal differentiation is not simply the study of regions, but the study of differences between regions, of the contrasts on the earth's surface and the reasons for these. Models can help here, but again geographers must decide which differences are important, significant and worth study. Then they must hypothesise reasons for the differences, reasons which are both relevant to man's use of the earth's surface and which can be tested and verified. A model-based approach could make regional geography much more relevant to modern problems by directing attention away from superficial differences and unproved explanations which have prevailed in the past.

One believes that the study of the evolution of landscapes is alive and well, and thriving in Berkeley. There are some hints of it flourishing in Germany and England (e.g., Reiner, 1970; Osborne and Wheeler, 1969) but this has had a less obvious, if extensive, effect on British geography. One presumes that models have been built to help in this kind of study, but the models are not as well publicised as those in general geography. Models such as sequent occupance, or reference to stages or cycles of development are obviously relevant here. Two important points can be made, even if some information is lacking. First, that models can be applied to this aim of geography just as to any other. One need not be misled by the types of models published in such books as *Models in Geography* into thinking that branches of general geography are the only studies in which models can be used. Second, following from this, that a different type of study, a different aim, demands a different type of model. Doubts were mentioned earlier about the dangers of borrowing models from other disciplines. If it is dangerous for general geographers to borrow models from science, it is virtually impossible for the landscape geographer to do so. He must make his own model, with, first, a clear idea of its purpose.

Reference to the effect of the landscape school on British geography will help to emphasise another point. Much British geography, ostensibly regional or general, has a large component in which the evolution of the landscape is described. This is the case when the location of industry is being 'explained' as much as when a region is being introduced. Now a full description of how a landscape came about, from uplifts in the Palaeozoic, through marine planations, periglaciations, deforestation by Bronze Age beaker people, description in Domesday Book, late enclosure, industrial revolution, to urban renewal, is neither explanation nor model. An explanation must contain some element which shows that this happened *of necessity*, while the model must contain both this functioning of necessity, and a normative formulation which makes it applicable to other cases.

Studies of man—land relationships fall into four groups: human ecology, the interaction between cultural and physical landscapes, man's adaptation to his environment in various regions and determinism itself. Of these four, the first is the most active at present while the last, although believed to be extinct, like the coelocanth floats up to the surface very occasionally. Perhaps few geographers regret the passing of the study of determinism, although it is still implicit in a surprising amount of work (see Minshull, 1970), yet this kind of study could benefit enormously from the use of models (Pahl, *MG*, ch. 7).

In the past, it was clear that those interested in determinism wanted to understand the influence of the physical environment on man's actions. This was one-sided, and the study of man—land interactions, influences in both directions, is a more sensible and worthwhile aim, but at least the aim was clear, the elements of the system had been identified, and it would have been possible to model the functioning and put forward hypothetical explanations. The most serious mistake in the history of the study of determinism was that Ratzel's hypotheses were accepted as verified by Semple. In an example such as this, one can see the great value of models: after forcing one to define one's aims, isolate the elements of the system and represent its functioning, they also force one into the position of being aware that an explanation is not true until it has been properly tested and verified. One of the most disturbing conclusions on reading Harvey is that now, *and only now*, is geography beginning to adopt the tools and methods effectively to study the aims it has claimed to pursue for centuries. Only now do models make it possible to study areal differentiation, the development of landscapes and the interaction of man and land, just at the time when geographers are losing interest in these aims.

Because models have not been applied in an obvious and conscious way to regional landscape and determinist geography, it is not surprising that their use is linked with general geography and with the study of spatial distributions. Equally, it is not surprising that some students who have seen the value of the use of models have been mystified to find that they were pursuing a different aim afterwards from that before. It is asserted again that the introduction of models does not necessitate a change in aims. However, if one still aims to study regions, landscapes or determinism, then it does necessitate the choosing and building of models to suit one's own aims and purposes. Without doubt, the worst aspect of the introduction of models into British geography is the diversion of many students along paths they never wanted to follow. This is partly the fault of the students and their teachers for not having their aims clear and their resolve firm, but it is mainly the fault of the advocates of models who are often, less obviously, also the advocates of the study of spatial distributions and geometric arrangements on the earth's surface.

Little more need be said here about the use of models in the study of spatial distributions because it is in this branch of geography that the main application has been made. The study of spatial distributions and relationships seems to be the main aim of geography at the moment and some claim, of course, that at last geographers are concentrating on that aspect which is unique to their discipline, and that therefore this is necessarily *the* overriding aim of geography. Others may hope for a re-application of the laws to be formulated in this new geography to a future study of regions in a more rigorous way than in the past, but it is certainly too early for this work to be possible yet. The distinction between the spatial distribution theme and the geometric theme is not completely clear, but Harvey (1969, p. 116), at least, thinks there is a possibility that the geometric theme in its turn will replace that of spatial distribution (see also Bunge). At this point in time one can conclude only that at least three different aims are going to be competing with each other in the future, and that preoccupation with models should not be allowed to cause any geographer to lose sight of his aims, as seems to have happened over the last fifteen years.

The purpose of this brief review of five aims of geography has been to make one point, that the major application of models to one of the aims should not obscure the fact that the others can still be pursued, and pursued more effectively with the proper use of models. However, it raises more questions which cannot be answered fully. The student may want to know which of these, or other, aims is the best to follow, but he must answer that question for himself. Others may ask which models are appropriate to regional geography, human ecology and the rest, but there is not space here to deal properly with all the models which exist. Even the models appropriate to the spatial distribution theme are not fully treated in books, and it is necessary to search the journals and to do some research to begin to grasp the complexity of each.

On this point, it is much more important to emphasise that models can and should be built to serve the particular purpose of the geographer, whatever he may be studying. Model building is very difficult; it will be achieved successfully by only a very few, but in the author's opinion it will be better for geography in the long run, and better for the education of individual students, if we see the slow accumulation of a few good, specific models, rather than an orgy of model borrowing and misapplication.

Another question is, should every model in geography be a spatial model, or have a spatial element? This has a close connection with the type of explanation to be given, which will be mentioned below. With the dominance of the study of spatial distributions, and the suggestion that the study of spatial arrangements is the only characteristic unique to geography which identifies it as a separate discipline, there is a strong temptation to answer yes to this question. However, very careful examination of published models will reveal that it is often completely impossible to identify any spatial element at all in a so-called 'geographical' model. This is not to say that a model must be graphical, or have a map, for at times a verbal or mathematical model without these elements can be translated into spatial terms. But between these and the very obvious examples of the demographic cycle, the rank-size rule, the circle of poverty, and so on, which have no spatial element at all, one finds a whole range of models which either have no spatial application, or have a spatial application in only the most tenuous way, for example in that a model of development refers to a country which has territorial extent.

Those who believe this question is redundant, who believe that obviously a model in geography does not need a spatial element, can point to many arguments. The bulk of regional geography contains no spatial analysis, the study of landscapes puts the emphasis on change in time rather than arrangements in space, geomorphology often studies landforms in isolation and puts the emphasis on process, population geography studies the age, sex and occupational structure of the people in each administrative unit, but not necessarily their specific spatial arrangement within those units.

Thus it is argued that many students will not expect 'geographical' models to have a spatial element because their concept of the nature of geography is distorted by the inclusion of such things as economic history, ethnology, geomorphology, demography and the like which do not demand spatial analysis (Minshull, 1970). Some aspects of climatology, contemporary economics, sociology and so on do have a spatial element, and one would suggest that when the geographer uses work from these closely connected disciplines, then, as a geographer and not a social scientist *manqué*, he concentrates on those geographical aspects which are defined by his particular aim in geography.

The spatial element in the study of different phenomena appears in different disguises and, of course, the mere cataloguing of which rocks, mountains, climate, farming, factories and towns are to be found within a certain patch of the earth's surface is no longer intellectually satisfying. Again the aim is to demonstrate that the phenomena are arranged on the earth's surface in a particular way, of necessity. The corries, arêtes, glacial trenches, moraines and outwash plains are arranged round a peaked mountain in a particular way of necessity because the glaciers behaved in a certain, uniform, predictable way. The wind systems, and consequently the climatic regions, show a logical pattern on the earth's surface, distorted as they may be by the shapes of the oceans, because they develop of necessity in this way, given the shape of the earth, the atmosphere, and the energy from the sun.

The climatic pattern, and the patterns of zonal soils and wild vegetation connected with them, show a unique arrangement on the earth's surface, the pattern tied to the equator and the poles, or, in other words, located in relation to these features. In contrast, like the arrangement of glacial features round a peak, many other patterns recur frequently on the earth's surface, but each time the pattern is tied to a fixed point: for example, the arrangement of fields round a farm or town, idealised in Von Thünen's model, of functional zones round a city centre, or of central places round the metropolis. From these examples it begins to be clear that the geographer can study several kinds of pattern, distribution and spatial arrangement on the earth's surface from one, worldwide, unique pattern, to thousands of small repeating patterns.

In the matter of distributions, the interest seems to have moved from unique locations to relative locations. Not so very long ago geographers studied individual factories and individual towns to see how and why they were tied to unique sites on the earth's surface. Often the relationship studied was that between man's structures and the rock, relief and soil of the physical environment. Nowadays, much more often, the geographer studies vast numbers of farms, factories and towns to see how and why they are spaced out. The interest is less in the site and the unique location, and much more in the situation, the relative location, in the relationships in space. Moreover, the geographer studies the relationship between different examples of man's structures as often as between these and the physical environment.

Other disciplines study the nature of the phenomena mentioned above, political, social and economic history deal with their development, and so it is argued that the characteristic work of the geographer is to study their necessary arrangements in space. Guelke (1971), in attacking models and the attempt to formulate laws in human geography, uses the argument that so many models do not contain a spatial element to try to reject models. One is inclined to agree with him that any geographical model should have a spatial element, but instead of rejecting all models because so many of those in use in geography clearly are not spatial, one must emphasise again the fact of the existence of several different aims within geography. In the pursuit of the aims of spatial relationships and the geometric aspects of the earth's surface, clearly, one would expect models with a spatial element. But in the case of the study of areal differentiation, landscape development and man—land relationships, process may well be the dominant theme, with arrangement in space being secondary or only implied.

Having given details of the three types of explanation, cause and effect, temporal, and functional, Harvey (1969, ch. 22) comes down in favour of the functional type of explanation in geography. Admittedly, his aim is to merge geography with science, and one hoped for a more obviously 'geographical' type of explanation at the end of such a long and exhaustive examination of explanation in geography, but it seems that this is the best available. This connects with the idea that models in geography should have a spatial element, in that any emphasis on the spatial arrangement of phenomena on the earth's surface will be cancelled out if the geographer then gives an historical type of explanation; again, the historical type of explanation, whether cause and effect or temporal, is so common in so much geography that the student may fail to grasp the point. The historian will tell us how the region, the industry, the landscape or the city came about, and, because of his professional training, will do it better than the geographer. It is incumbent on the man who rightly calls himself a geographer to explain to us how the phenomena are now arranged, and now function, on the earth's surface. The one snag when Harvey recommends the functional type of explanation as the best available to geographers is that some of the phenomena studied by them at present (one thinks of rocks) do not permit a functional explanation.

While it is easier to take a stand at one extreme or the other, neither Guelke nor Chorley and Haggett are entirely convincing, and a middle way seems necessary. Guelke is most critical of models and sees them as irrelevant to his view of the nature of geography. Chorley and Haggett, having a different view of the nature of geography, give the impression that in models geographers at last have the potential answer to all their problems. It has been suggested above, however, that some models can be developed which are relevant to whatever aim of geography an individual decides to pursue. In addition, it must now be stressed that however useful models prove to be, they are no more the complete solution to geography's problems than any other technique or device in the past.

The use of models in research into any branch of geography should prove more effective than techniques applied in the past. To state this is not to demand a rejection of any other techniques which have proved useful, nor is it to demand that any individual geographer change his aims. The use of models in teaching, in exposition, at any level, is a different matter at present. Models are so effective, the functional explanation of a system is so much more effective in teaching than the dictation of a mass of unconnected facts that the use of models in this way, particularly in human geography, will prove to be most important in the long run. 'In the long run' must be stressed, because, whatever impression certain books may give, there are not sufficient *verified* models to be used for teaching at the moment. Teaching by means of models can be so effective that there is a grave danger of teachers and lecturers using untested, unverified hypotheses as if they were established, universal truths, especially with the encouragement they are being given in this direction. The author is convinced of the long-term effectiveness of models, but would not be surprised to see them become as embarrassing a skeleton in the cupboard as determinism through overhasty adoption and uninstructed, unquestioning use.

To advocate the use of models is not necessarily to insist on the study of general geography, of spatial distribution, or on the aim to formulate general laws. The most significant result of the development of geographical models in the long term will be the establishment of genuine principles as distinct from superficial generalisations. But one must insist that neither the construction of models nor the formulation of principles is the end purpose of geography. One insists that the end purpose is understanding of the earth's surface, and, if necessary, particular parts of the earth's surface. It is perfectly possible, when, eventually, sufficient models have been tested and accepted, to re-apply the principles so gained to a more enlightened investigation of particular regions.

Among the many other reasons for saying this, including an abiding interest in regional geography, is an important educational one which has implications for everything that an educated person does, far beyond the limits of geography. Time and again, students will mistake a reiteration of general principles for a particular explanation. Granted we are at the stage where what models we possess still have to be tested and modified, there is still the mistake that the general, normative explanation is sufficient and necessary explanation of a particular case, even when the particular case blatantly contains something not covered in the general model.

This is where geography *is* different from natural science. This is where the exceptions, the discrepancies, the unique, cannot be dismissed as resulting from the fact that the earth's surface is not an isotropic plain. Geographers once professed to be interested in the earth's surface, and while one takes Guelke's point that they must not end up just studying the *discrepancies* between normative model and actual case, one insists that a geographer must end up by studying the whole of the particular case on the earth's surface, normative element, special element and all. The full explanation of arrangements on the earth's surface may well demand the application of both general laws and seemingly special pleading until, a few thousand years from now, men have the knowledge and techniques to see what appear to be special cases as part of a more general system. As Guelke states, the new geographers 'have achieved internal consistency while losing their grip on reality'.

The introduction of models, to geography, among other useful effects, can force some of us, teachers, geographers and students alike, to stop and think just what it is we are trying to do. The choice is not just between human and physical, regional and general geography, nor is it just among areal differentiation, landscape evolution, human ecology, spatial distribution and the recognition of geometrical patterns. The choice can also be between objective academic study and political practical planning. It can be between aiming 'to build . . . coherent theories . . . to search for order in a complex world' (Cooke and Johnson, 1969, p. 81), and aiming to comprehend those parts of the world which one has seen. The teacher may aim to turn out more geographers, or to turn out educated laymen who have been subjected to rigorous scholarship, where the pursuit of truth, logic and an analytical approach is more important than particular facts and professional expertise.

Guelke asks whether we aim at truth or usefulness, a pertinent question with geography in its present state. So many simulation models used for prediction and planning are black box models, and they can be useful in their short-term practical application while presenting a completely untrue hypothesis of the nature of the system. Moreover, so many geographers seem to have an inferiority complex which manifests itself in an urge to plan, to be useful in a way that the masses will understand. Guelke insists that in geography the aim must be truth, a true understanding of the nature of things, rather than short-term usefulness. In this context he attacks Haggett's (1965, p. 23) recommendations that models are inevitable, stimulating and economical, insisting that they are not inevitable if they do not get us nearer to the truth, and that stimulation is irrelevant, again if it does not lead toward truth. Education must lead students to be analytical, and anxious to understand, to get at the truth. If models in geography cannot also serve this purpose, then geography will inevitably become a mere stimulating pastime like bridge or bingo.

This introduction to the use of models in geography, to the much lengthier and more complex works on the matter listed in the Bibliography, while clearly recommending an understanding of models, must in no way minimise their dangers and complexities. In the past, I have known several very intelligent students who were really never attracted to geography or who, at some stage, changed to some other discipline because they felt that geography was not sufficiently intellectually demanding, and, thereby, ultimately incapable of being satisfying. Therefore one must give greater credit to those able people who have worked all the harder to improve the standard of geographical research, not for prestige in comparison with other disciplines, but for the intellectual benefit of all those who are attracted to geography because of the things it studies and how it studies them. Much of the credit must go to those who have been concerned with techniques, and with quantitative techniques and models in particular.

The introduction of models into geography is a decided advance, but it has also introduced complications of which many people are only dimly aware. Moreover, the real advantages will be gained only over a very long time, as models are built, tested, rejected, rebuilt, tested, modified, tested and tested again. This book is only a very short, very simple introduction to a very complicated world. Models may make geography more accurate and more truthful, but they will make it much more difficult. The correct representation of systems, an understanding of their functions, and the hypothesis of possible explanations demands the best in training, patience, intelligence and insight. The most encouraging aspect of the new model-based paradigm is that now, geography, to succeed, will demand as much intelligence, patience and skill as any other profession or discipline.

Bibliography

ACKOFF, R. L., GUPTA, S. K., and MINAS, J. S. (1962) *Scientific Method: optimizing applied to research decisions,* Wiley.

AMBROSE, P. (1969) *Analytical Human Geography,* Longmans.

BALCHIN, W. G. V. (1970) *Geography: an outline for the intending student,* Routledge & Kegan Paul.

BERRY, B. J. L. (1967) *A Geography of Market Centres and Retail Distribution,* Prentice-Hall.

BERRY, B. J. L. and MARBLE, D. F., eds. (1968) *Spatial Analysis,* Prentice-Hall.

BIRD, J. (1971) *Seaports and Seaport Terminals,* Hutchinson University Library.

BOURNE, L. S., ed. (1971) *Internal Structure of the City,* Oxford University Press.

BROEK, J. O. M. (1965) *Geography, its Scope and Spirit,* Merrill.

BUNGE, W. (1962) *Theoretical Geography,* Lund Studies in Geography Series C, General and Mathematical Geography 1, Lund (Sweden), Gleerup.

CHISHOLM, M. (1966) *Rural Settlement and Land Use,* Hutchinson University Library.

CHISHOLM, M. (1967) 'General systems theory and geography', *Transactions of the Institute of British Geographers,* No. 42.

CHISHOLM, M. (1970) *Geography and Economics,* Bell.

CHORLEY, R. J. and HAGGETT, P., eds. (1967) *Models in*

Geography (Madingley Lectures), Methuen [cited as *MG*].
1. P. Haggett and R. J. Chorley, Models, paradigms and the new geography.
2. F. H. George, The use of models in science.
3. R. J. Chorley, Models in geomorphology.
4. R. G. Barry, Models in Meteorology and climatology.
5. R. J. More, Hydrological models and geography.
6. E. A. Wrigley, Demographic models and geography.
7. R. E. Pahl, Sociological models and geography.
8. D. E. Keeble, Models of economic development.
9. B. Garner, Models of urban geography and settlement.
10. F. E. I. Hamilton, Models of industrial location.
11. J. D. Henshall, Models of industrial activity.
12. David Grigg, Regions, models and classes.
13. D. R. Stoddart, Organism and ecosystem as geographical models.
14. D. Harvey, Models of the evolution of spatial patterns.
15. C. Board, Maps as models.
16. M. A. Morgan, Hardware models in geography.
17. S. G. Harries, Models of geographical teaching.

CHORLEY, R. J. and HAGGETT, P. (1970) *Frontiers in Geographical Teaching*, Methuen.

CHORLEY, R. J. and HAGGETT, P. (1969) *Network Analysis*, Arnold.

COHEN, S. B. ed. (1967) *Problems and Trends in American Geography*, Basic Books.

COLE, J. P. and KING, C. A. M. (1968) *Quantitative Geography*, Wiley.

COOKE, R. U. and JOHNSON, J. H. (1969) *Trends in Geography*, Pergamon.

CHRISTALLER, W. (1933) *Die Zentralen Orte in Süddeutschland*, Jena.

DAVIS, W. M. (1910) *Geographical Essays*, Boston.

DEMANGEON, A. (1933) 'Une carte de l'habitat', *Annals de Géographie*, 42.

ESTALL, R. (1972) *A Modern Geography of the United States*, Penguin (Pelican).

EYRE, S. R. and JONES, G. R. (1966) *Geography as Human Ecology*, Arnold.

FORDE, C. DARYLL (1934) *Habitat, Economy and Society*, Methuen.

FREEMAN, T. W. (1969) *Ireland*, Methuen.

GREGORY, S. (1973) *Statistical Methods and the Geographer*, 3rd edn, Longman.

GRIGG, D. (1969) 'The agricultural regions of the world', *Economic Geography*, 49, No. 2.

GUELKE, L. (1971) 'Problems of scientific explanation in geography', *The Canadian Geographer*, 15, No. 1.

HAGGETT, P. (1965) *Locational Analysis in Human Geography*, Arnold.

HARTSHORNE, R. (1939) 'The nature of geography', *Annals of the Association of American Geographers*, 29.

HARVEY, D. (1969) *Explanation in Geography*, Arnold.

HOLMES, A. (1965) *Principles of Physical Geology*, Nelson.

HOUSTON, J. M. (1970) *A Social Geography of Europe*, Duckworth.

JONES, EMRYS (1964) *Human Geography*, Chatto & Windus.

JOHNSON, J. H. (1967) *Urban Geography*, Pergamon.

KING, L. J. (1969) *Statistical Analysis in Geography*, Prentice-Hall.

KING, C. A. M. (1970) 'Geography: an outline for the intending student', in Balchin (1970).

McCARTY, H. H. and LINDBERG, J. B. (1966) *A Preface to Economic Geography*, Prentice-Hall.

MARTIN, A. F. (1951) 'The necessity for determinism', *Trans. Inst. British Geographers*, No. 17.

MAYER, H. M. and KOHN, C. F. eds. (1959) *Readings in Urban Geography*, University of Chicago Press.

MINSHULL, R. M. (1967) *Regional Geography: theory and practice*, Hutchinson University Library.

MINSHULL, R. M. (1970) *The Changing Nature of Geography*, Hutchinson University Library.

MONEY, D. C. (1965) *Climate, Soils and Vegetation*, University Tutorial Press.

MORRILL, R. L. (1970) *The Spatial Organisation of Society*, Wadsworth.

MOUNTJOY, A. B. (1963) *Industrialisation and Underdeveloped Countries*, Hutchinson University Library.

OPEN UNIVERSITY (1970) *Understanding Society: readings in the social sciences*, Macmillan.

OPEN UNIVERSITY (1970) D100,S 04, CN, *Models in Geography*.

OSBORNE, R. H. and WHEELER, P. T. (1969) *Rural Studies in the North East Netherlands*, Geography Field Group, University of Nottingham, Regional Study, No. 14.

PRED, A. (1964) 'The intrametropolitan location of American manufacturing', *Annals of the Association of American Geographers*, 54.

REINER, E. (1970) 'Das Land System Konzept und das Muster in der Luftbild Interpretation', *Bildmessungen und Luftbildwesen*, 5.

RUTHERFORD, J., LOGAN, M. I., and MISSEN, G. J. (1968) *New Viewpoints in Economic Geography*, Harrap.

SACK, R. D. (1972) 'Geography, geometry and explanation', *Annals of the Association of American Geographers*, 61, No. 1.

STRAW, A. (1971) 'Systems analysis and physical geography', Bishop Grosseteste College, Lincoln.

TAAFFE, E. J. (1970) *Geography*, Prentice-Hall.

TAYLOR, G. (1949) *Urban Geography*, Methuen (2nd edn, 1968).

THEAKSTONE, W. H. and HARRISON, C. (1970) *The Analysis of Geographical Data*, Heinemann Educational.

THÜNEN, J. H. VON (1826) *Der isolierte Staat in Beziehung auf Landwirtschaft und Nationalökonomie*, Hamburg.

TIDSWELL, W. V. and BARKER, S. M. (1971) *Quantitative Methods*, University Tutorial Press.

TOYNE, P. and NEWBY, P. (1971) *Techniques in Human Geography*, Macmillan.

WILSON, A. (1970) *War Gaming*, Penguin (Pelican).

WOOLDRIDGE, S. W. and MORGAN, R. S. (1959) *The Physical Basis of Geography*, Longmans.

YEATES, M. H. (1968) *An Introduction to Quantitative Analysis in Economic Geography*, McGraw-Hill.

Index